# THE CHEMISTRY OF PAPER

# RSC Paperbacks

RSC Paperbacks are a series of inexpensive texts suitable for teachers and students and give a clear, readable introduction to selected topics in chemistry. They should also appeal to the general chemist. For further information on selected titles contact:

Sales and Promotion Department
The Royal Society of Chemistry
Thomas Graham House
The Science Park
Milton Road
Cambridge CB4 4WF, UK
Telephone: +44 (0) 1223 420066

## Titles Available

**Water** *by Felix Franks*
**Analysis – What Analytical Chemists Do** *by Julian Tyson*
**Basic Principles of Colloid Science** *by D.H. Everett*
**Food – The Chemistry of Its Components (Third Edition)**
*by T.P Coultate*
**The Chemistry of Polymers** *by J.W. Nicholson*
**Vitamin C – Its Chemistry and Biochemistry**
*by M.B. Davies, J. Austin, and D.A. Partridge*
**The Chemistry and Physics of Coatings**
*edited by A.R. Marrion*
**Ion Exchange: Theory and Practice, Second Edition**
*by C.E. Harland*
**Trace Element Medicine and Chelation Therapy**
*by D.M. Taylor and D.R. Williams*
**Archaeological Chemistry** *by A.M. Pollard and C. Heron*
**The Chemistry of Paper** *by J.C. Roberts*

## How to Obtain RSC Paperbacks

Existing titles may be obtained from the address below. Future titles may be obtained immediately on publication by placing a standing order for RSC Paperbacks. All orders should be addressed to:

The Royal Society of Chemistry
Turpin Distribution Services Limited
Blackhorse Road
Letchworth
Herts SG6 1HN, UK
Telephone: +44 (0) 1462 672555

RSC Paperbacks

# THE CHEMISTRY OF PAPER

## J.C. ROBERTS

*Department of Paper Science, UMIST, Manchester*

THE ROYAL
SOCIETY OF
CHEMISTRY
Information
Services

ISBN 0-85404-518-X

A catalogue record for this book is available from the British Library

© The Royal Society of Chemistry 1996

Published by The Royal Society of Chemistry, Thomas Graham House,
Science Park, Milton Road, Cambridge CB4 4WF, UK

Typeset by Keytec Typesetting Ltd, Bridport, Dorset
Printed and Bound By Athenaeum Press Ltd, Gateshead, Tyne & Wear

# Preface

For what is usually thought of as an essentially mechanical process, paper manufacture involves a surprisingly large amount of chemistry. From the conversion of wood into pulp to the formation of the final sheet, chemical principles are important. The delignification of a suitable plant source, usually wood, is a chemically heterogeneous process which is performed at elevated temperature and pressure. The lignin of the lignified plant tissue is solubilised from the wood matrix, thereby liberating the component fibres as a cellulose-rich pulp from which the paper or board will be made. The pulp is then often bleached, and the chemistry of this process has changed considerably over the past decade because of the adverse environmental impact of chlorine-based bleaching systems. The component fibres are finally formed from a wet suspension into a bonded network sheet structure whose mechanical strength is provided by both the fibres themselves and by extensive inter-fibre hydrogen bonding. A great deal of chemistry is involved in this formation process, especially as it is often necessary for the sheet to be modified in order to give it properties which are appropriate to its end use.

The chemistry is both wide ranging and interesting. It involves carbohydrate chemistry, the chemistry of inorganic pigments, organic resins—both natural and synthetic—and many other organic and polymeric additives. The sheet formation process also involves a considerable amount of colloid and surface chemistry. Polymer chemistry and environmental and analytical chemistry also play an important part.

My primary objective has been to provide an introduction to the most important parts of the process of making paper in which chemistry plays a role, and I have attempted to deal with the chemistry of the process in more or less chronological order. Because of the complexity of many of the processes, and also the abundance of available literature, the subjects are necessarily covered somewhat briefly. However, I have attempted, where possible, to emphasise important principles, and the reader is directed to other recommended reading for a more comprehensive coverage. This book should be suitable to anyone who has a reasonable knowledge of chemistry to around degree level but who requires an introduction to the chemistry of paper manufacture.

# Dedication

To Francis John and Christopher Leslie.

# Contents

# Acknowledgements

My grateful thanks are extended to Guomei Peng who gave me enormous help with collecting data and constructing tables and graphs; to Dr. Chris Wilkins for contributing the photomicrographs; and to Huguette Chatterton, Pam Kirk and Rachel Parker who, at various times, converted my rather confused dictations into a workable format. I would also like to thank Professor Kit Dodson (University of Toronto), Dr. Derek Priest (UMIST) and the staff of Weyerhaeuser Technical Centre in Seattle for many valuable discussions. I would also like to thank UMIST for granting me a year of study leave to prepare this book, and my colleagues in the Paper Science Department of UMIST for shouldering many of my normal responsibilities during that year. Last but not least I am indebted to my wife, Lesley, and my children for their patience in bearing with my absence for many hours during the preparation of the manuscript.

*Chapter 1*

# An Introduction to Paper

## INTRODUCTION

Paper has been an essential part of our civilisation for at least two thousand years and, perhaps because of our familiarity with it, we do not tend to think of it as a particularly complex material. However nothing could be further from the truth. It is derived from plant sources and therefore has both morphological complexity and physical and chemical complexity. Even our understanding of its load–elongation behaviour, which might be expected to be relatively simple, is still far from complete. The production process itself is also highly sophisticated, involving what is in essence a high-speed filtration process yielding a weak wet fibrous network. This wet web, despite its weakness, must then be pulled continuously through the pressing and drying sections of the paper machine to the reel at speeds which these days approach $60 \, km \, h^{-1}$, during which the web undergoes some extension. To avoid frequent breaks, and to obtain good product uniformity therefore requires some of the most advanced control engineering technology available today.

This opening chapter is a brief introduction to the nature of paper, its history and to its modern day use.

## DEFINITION OF PAPER

When we think of paper we think of it primarily as a writing and printing medium, and then perhaps as a wrapping and packaging material. However, because many other products—for example, tissue, board, filtration media, surgical wrap, *etc.*—are made by essentially the same process, a broader definition is more appropriate. For the purpose of this text therefore, paper will be defined in

1

terms of its method of production, that is a sheet material made up of a network of natural cellulosic fibres which have been deposited from an aqueous suspension. The product which is obtained is a network of interlocking fibres with an approximately layered structure about 30–300 μm thick. The width of an individual fibre is in the range 10 to 50 μm, and a sheet of writing paper of 100 μm thickness would therefore be expected to be about 5 to 10 fibres thick (Figure 1.1).

The precise time and place at which paper was introduced into our civilisation is not known with any certainty. Before 700 BC, animals skins were certainly the medium of written communication, but these were displaced by papyrus by the Egyptians at some time around 600 BC. Papyrus, although derived from a plant source, is not strictly paper as defined above, as it is made by separating and spreading the pellicles of the aquatic papyrus plant on to a flat surface sprinkled with water rather than by depositing a network of fibres from an aqueous suspension. Various forms of parchment, which are close relatives of papyrus, were used by the Greeks and Chinese for the next 800 years, but it was not until around 200 AD that the Chinese introduced the art of making paper by reducing fibrous matter to a pulp in water and then forming it as a network. The Chinese are thus usually credited with the invention of modern paper manufacture.

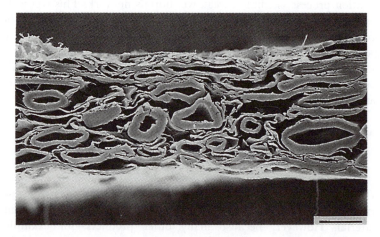

**Figure 1.1** *Scanning electron photomicrograph of a cross section of a national newspaper comprising 90% spruce and 10% pine thermomechanical pulp (TMP) fibres (45 g m⁻² and ~8 fibres thick). Scale bar = 25 μm.*

The fibres from which paper is made are the structural cells of plants, and paper could therefore be made, in principle, from a wide variety of plant sources. In practice, the sources are limited by factors such as availability, crop yield per hectare, and quality of the fibre. In the late nineteenth and early twentieth centuries cotton in the form of rags was the main fibre source, and the pioneer paper-making factories grew up around the sites of the textile manufacturing industry. Since the early part of the twentieth century, as the demand for paper grew and the waste from the textile industry was no longer able to satisfy the demand, wood became increasingly used, so that now over 90% of virgin fibre (that is excluding any recycled fibre) is derived from wood.

## PRODUCTION AND CONSUMPTION

The annual world production of paper and board is around 250 million metric tonnes, and well over half of this is produced in the US and EEC countries. A mere 1.2 million tonnes is produced in the whole of Africa. It is also consumed almost totally by the developed world and the per capita consumption of paper and board products varies hugely throughout the world (Table 1.1).

In addition to fibre obtained directly from plant sources by chemical or mechanical treatment (virgin fibre), recycled fibre is also used and to an increasing extent for paper and board production. A breakdown of world fibre usage is given in Table 1.2 and the subject

**Table 1.1** *Annual per capita consumption (1991) of paper and board products in various regions world-wide.*
(Source: 1993 USA Pulp and Paper Fact Book).

| Region | Annual per capita consumption of paper and board/kg |
|---|---|
| USA and Canada | 294 |
| Japan | 248 |
| Nordic Countries | 213 |
| EEC | 156 |
| Australasia | 126 |
| S. Africa | 43 |
| Eastern Europe | 29 |
| S. America | 28 |
| Asia | 21 |
| Africa (excl. S. Africa) | 3 |

**Table 1.2**  *The source of fibre world-wide for paper and board production.*
(Source: Pulp and Paper International, 'International
Fact and Price Book', 1994).

| Year | World pulp production from virgin fibre (million tonnes) | World paper and board consumption (million tonnes) | Recycled fibre usage (million tonnes) | Recycled fibre usage (%) |
|------|------|------|------|------|
| 1992 | 163.5 | 245.7 | 82.2 | 33.5 |
| 1991 | 162.6 | 239.4 | 76.8 | 32.1 |
| 1990 | 162.6 | 238.1 | 75.5 | 31.2 |
| 1989 | 164.1 | 231.7 | 67.6 | 29.2 |
| 1988 | 161.7 | 225.3 | 63.6 | 28.2 |
| 1987 | 154.3 | 214.3 | 60.0 | 28.0 |
| 1983 | 131.1 | 176.9 | 45.8 | 25.9 |
| 1978 | 122.2 | 157.6 | 35.4 | 22.5 |

of paper recycling and its chemistry is discussed more fully in
Chapter 9.

Recycled fibre now accounts for over a third of all fibrous raw
material and, over the past few years, its use has steadily increased
whilst that of virgin pulp has remained fairly constant. The extent to
which recycled fibre is used varies greatly from country to country.
In Europe, where there is a fibre deficiency, it accounts for over half
of the total fibrous raw material whereas in North America and
Canada, where wood is plentiful, recycling levels are much lower.
There is still scope therefore to increase further the use of recycled
fibre and, as the consumer is increasingly demanding it in paper and
board products, the upward trend is expected to continue for some
time yet. Recycled fibre is not distributed uniformly through all
grades of products; some grades—for example many types of
board—use 100% whereas others, such as speciality grades and
some high quality writing grades use none at all. This subject is
discussed more fully in Chapter 9.

## FIBRE SOURCES

Although the amount of recycling could still be increased, there is
almost certainly an ultimate limit to the extent to which recycled
fibre can be used, and it is difficult to foresee a totally 'closed fibre'

industry in which no new fibre is introduced. Most of the newly introduced fibre will also probably continue to be derived from wood, although annual crops can be expected to play an increasingly important role.

Approximately 30% of the earth's land surface is forested, and around half of this is harvested commercially. Over 80% of the wood for all industrial uses comes from the forests of North America, Europe and what was formerly the Soviet Union. Approximately two thirds of this is either sawn or peeled. Paper is generally made either from logs that are unsuitable for sawing or peeling or from residues arising from these processes.

Both hardwoods and softwoods are used for making paper and they have very different fibre morphologies and thus very different paper-making properties. The fibres of softwoods are longer and stronger than those of hardwoods and they make up the bulk of paper-making fibre world-wide (Table 1.3). However, because they easily form macroscopic flocs of entangled fibres during the sheet forming process, they tend to produce a sheet with a relatively non-uniform mass distribution and hence a poorer quality of appearance (this is known by paper technologists as formation). It is common therefore to use blends of softwood and hardwood fibres to give an appropriate compromise between strength and formation. The characteristics of hardwood and softwood fibres are discussed at greater length in Chapter 2.

Non-woody fibre, although relatively small in volume is nevertheless important, particularly in the developing world where the use of indigenous raw materials can substantially reduce the amount of foreign exchange spent on importing costly wood pulp. The main sources of these fibres are bagasse, bamboo, jute, ramie, hemp, flax and cotton, and also various grasses and straws, such as esparto,

**Table 1.3**  *World-wide hardwood, softwood and non-wood pulp production (1988).*

(Source: S. Dillen and H. Norstrom. *Pulp and Paper International*, 1990, **32** (10), p. 61–65).

|  | Million tonnes | % |
|---|---|---|
| Softwood pulp | 99.2 | 62 |
| Hardwood pulp | 41.6 | 26 |
| Non-wood pulp | 19.2 | 12 |
| Total | 160 | 100 |

wheat, barley or rice. Their main advantage over wood is that they can frequently be grown in areas which will not support trees, and in limited rainfall in low quality soil. In general, they produce an annual crop with a higher yield than wood. For example, straw can be produced at yields as high as 20 metric tons per hectare, which is considerably greater than the annual growth of most tree species. Non-woody plants can also be harvested relatively quickly—usually one or two years after planting—whereas trees require ten to twenty years to reach sufficient maturity.

The paper-making properties of all of these fibres are quite different from each other and also from wood. This is mostly due to the differing morphology and to some extent the differing chemistry of the fibre cells. The photomicrograph (Figure 1.2), shows a comparison between various non-woody fibre types.

## PRODUCT TYPES

Just over 40% of all the paper which is produced throughout the world is used for communication purposes (newsprint and printing and writing), and over 50% is used for packaging and tissue (Figure 1.3). The remainder is used in rather specialised applications such as filtration media, tea bags and electrical insulation in transformers.

Paper is classified in terms of its weight per unit area (basis weight or grammage). Tissue grades are generally in the range $10-40 \, \mathrm{g \, m^{-2}}$, newsprint around $40-50 \, \mathrm{g \, m^{-2}}$, printing and writing grades around $60-90 \, \mathrm{g \, m^{-2}}$, and boards are usually in excess of $100 \, \mathrm{g \, m^{-2}}$. Because of the need to obtain specific characteristics in the final product, for example water absorbency or wet strength, there is a great difference in the chemistry and method of production of these grades.

## CHEMICAL COMPOSITION OF PAPER

As paper is obtained from fibres which were, before chemical and mechanical treatment, the cells of land plants, it does not have a fixed chemical composition but one which is largely pre-determined by the fibre source. The cells of land plants are mostly composed of carbohydrate polymers (polysaccharides) which are impregnated to varying degrees, with lignin—a complex aromatic polymer the amount of which generally increases with the age of the plant and which is biosynthesised during the process of lignification. These

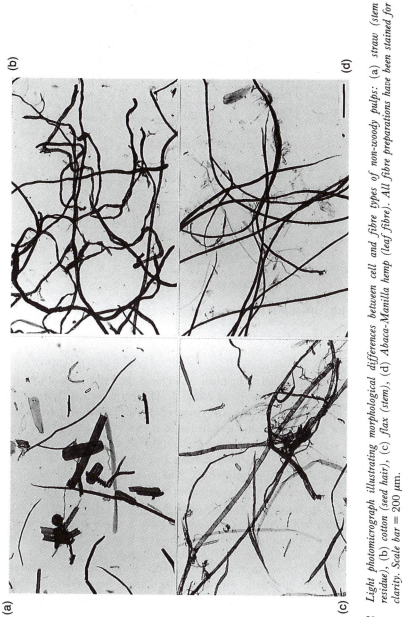

**Figure 1.2** *Light photomicrograph illustrating morphological differences between cell and fibre types of non-woody pulps: (a) straw (stem residue), (b) cotton (seed hair), (c) flax (stem), (d) Abaca-Manilla hemp (leaf fibre). All fibre preparations have been stained for clarity. Scale bar = 200 µm.*

| Grade | Million Tonnes |
|---|---|
| Newsprint | 31.8 |
| Printing & Writing | 70.2 |
| Tissue | 14.2 |
| Packaging | 112.8 |
| Other | 17.5 |

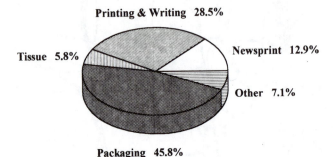

**Figure 1.3** *World-wide distribution of paper and board product types (1992).* (Source: Pulp and Paper International, 'International Fact and Price Book', 1994).

components and their chemical structures and functions are discussed more fully in Chapters 2 and 4. The carbohydrate part of the cell is dominated by the structural polysaccharide cellulose, but there are also other polysaccharides of a non-structural nature and with a very much lower molecular weight which are known, somewhat misleadingly, as hemicelluloses and which play an important part in pulp and paper properties. The term hemicellulose seems to imply some relationship to cellulose and, at one time, they were thought to be biosynthetic precursors of cellulose. However, it is now well established that these polysaccharides are not involved in the biosynthesis of cellulose, but are a discrete group of polymers with their own specific function in the plant cell wall.

In addition to these main components there are also relatively small amounts of organic extractives and trace inorganic materials. The approximate distribution of the three main groups of components together with other trace materials is given in Table 1.4.

The overall composition of plant fibre cells in terms of carbon, hydrogen and oxygen is variable and dependent on the degree of lignification. For wood it is approximately 50% carbon, 6% hy-

**Table 1.4** *Distribution of main chemical components of wood.*

|  | Cellulose (%) | Hemicelluloses (%) | Lignin (%) | Extractives and trace materials (%) |
|---|---|---|---|---|
| Softwood | 40–45 | 20 | 25–35 | < 10 |
| Hardwood | 40–45 | 15–35 | 17–25 | < 10 |

drogen and 44% oxygen. Carbohydrates, because they all have more or less the same elemental composition of $(CH_2O)_n$, have a more or less uniform carbon content of around 40%. Lignin, on the other hand, is an aromatic polymer with the approximate composition $C_{10}H_{11}O_4$ and therefore has a much higher average carbon content of about 60–65% (Table 1.5).

## CONVERSION OF NATURAL FIBRES INTO PAPER

Paper can be made from fibre cells in their more or less unmodified form, by simple mechanical disintegration to disperse them in water, and then forming them into a web by the process described in Chapter 5. This process of mechanical pulping is suitable only for products with a short life span—because the lignin (which is not removed) discolours in sunlight as a result of photochemically catalysed oxidation processes, and the paper becomes yellow and brittle. The use of lignin-containing fibres is therefore restricted to products such as newsprint and disposable light-weight coated paper. For higher quality papers which are required to have a longer lifetime, it is necessary to remove the lignin by a chemical pulping

**Table 1.5** *Approximate C, H, O content of lignin from spruce and beech.* (Source: 'Lignin Biodegradation: Microbiology, Chemistry and Potential Applications' eds. J. Kirk Kent, T. Higuchi and H. M. Chang, Vol. 1, CRC Press, Florida, 1980).

|  | Formula | C (%) | H (%) | O (%) |
|---|---|---|---|---|
| Spruce lignin (softwood) | $C_{9.92}H_{10.68}O_{3.32}$ | 65.1 | 5.8 | 29.1 |
| Beech lignin (hardwood) | $C_{10.39}H_{11.66}O_{3.92}$ | 62.6 | 5.9 | 31.5 |

process. This involves a high temperature and pressure reaction in which the lignin is solubilised under aqueous alkaline, neutral or acidic conditions. Non-aqueous solvent pulping procedures have also been developed but are not yet in full commercial use. The chemical removal of lignin produces a brown pulp, the colour of which is mostly due to chromophores associated with small amounts of residual lignin. It is therefore often followed by a bleaching operation which, in the past, has been almost exclusively chlorine-based but, as a result of environmental pressures, is being superseded by other methods. Chemical delignification and subsequent bleaching are discussed more fully in Chapter 3. Such fibres will be used in high quality printing and writing grades, and in high added-value speciality applications.

*Chapter 2*

# The Material of Paper

## INTRODUCTION

Unlike most chemical raw materials, the fibres which are used for paper making are produced not synthetically but biosynthetically as plant cells. The paper maker therefore, apart from using crop selection and strategies for growth and harvesting, has little control over fibre shape and chemical composition. As these have a profound influence upon the subsequent chemistry of the paper-making process, and also upon the physical and mechanical properties of the end product, it is important to understand something of the morphology, structure and chemical composition of paper-making fibres.

## FIBRE MORPHOLOGY AND WOOD CELL STRUCTURE

Plant cell walls may have shapes varying from spherical to cylindrical, and sizes varying from under 1 mm to several centimetres. In higher plants, two types of functional cell walls can be distinguished. These are the primary cell wall, which surround the growing cell, and the secondary cell wall, which is laid down when growth has ceased. The cell wall is a complex composite material and contains both structural and non-structural components. These components are mainly polysaccharides, although lignin and proteins also play an important part. The structural component is usually partly crystalline, and exists in the form of microfibrils. The most common of these is cellulose, which is a linear $\beta$-1,4-linked polysaccharide of $\beta$-D-glucopyranose, the molecular and crystal structure of which is discussed more fully in Chapter 4. Some algae contain structural polysaccharides composed of mannose and xylose units, but these have no industrial importance in paper manufacture. The non-structural polysaccharides are chemically more complex, and their function in the plant cell wall is still poorly understood.

The cellulose in wood and other species is present as microfibrils which are arranged in parallel lamellae and which occur in a number of orientations with respect to the cell axis. In various species of green algae, there are two main microfibrillar orientations, which are arranged in a shallow and a steep helix running round the vesicles. In the higher plants such as wood, the cell wall is made of co-axial layers of cellulose microfibrils, embedded in an amorphous matrix of hemicellulose and, in the later stages of the growing cycle, of lignin.

Predominantly two types of wood—hardwood and softwood—are used for paper making. Softwoods are used more frequently because of their relatively long fibre length but hardwoods, although shorter in fibre length, play an important role in assisting the formation of the sheet. The relative amounts of each of these types of wood used for paper making world-wide are given in Table 1.3 (Chapter 1).

The terms hardwood and softwood are not well-defined and do not, as might be imagined, reflect the physical properties of the timber. Some hardwoods are relatively soft, and some softwoods may be relatively hard. Strictly, the correct definition is a botanical one. Gymnosperms (softwoods) are species in which the seed is exposed and which are evolutionary older and therefore simpler in structure than the angiosperms (hardwoods), which have their seeds enclosed. The terms conifer (softwood, gymnosperms) and broadleaf (hardwood, angiosperms) are also often used in place of these more precise botanical definitions. The anatomical structures of hardwoods and softwoods are quite different as demonstrated by the photomicrographs in Figure 2.1. The dominant cells of each type of wood are also very different (Figure 2.2).

In softwoods, the main cell type is the tracheid, which is often mistakenly referred to as a fibre. Tracheids constitute over 90% of the volume of most softwoods, and are the principal paper-making cells of softwoods. Their average length is usually between 2 and 4 mm, with a length:width ratio (aspect ratio) often in excess of 100 to 1, but there is a wide distribution of tracheid lengths, and it is possible for some to be as short as 1 mm and for others to be as long as 5 mm (Table 2.1). The lumen, or central cavity, is several times wider than the cell wall thickness. There is also a difference between spring wood (*i.e.* cells synthesised in the early part of the annual

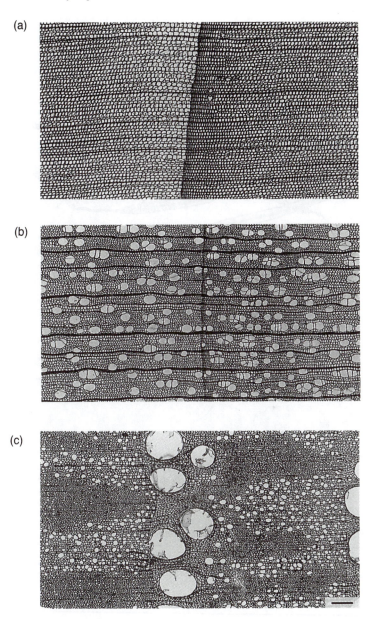

**Figure 2.1** *Light photomicrographs of wood cross-sections illustrating different anatomical features of softwood and hardwood: (a) pine (softwood), (b) birch (diffuse porous hardwood), (c) oak (ring porous hardwood). Scale bar = 200 μm.*

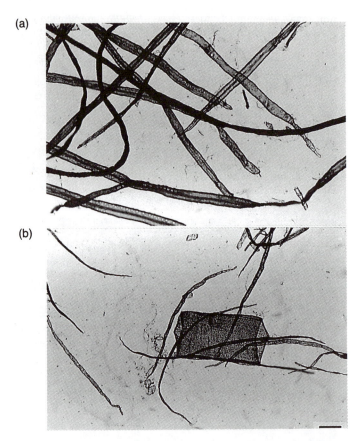

(a)

(b)

**Figure 2.2**  *Light photomicrographs of fibre preparations illustrating the morphological differences between softwood and hardwood commercial pulps: (a) bleached sulfate pine (softwood), (b) bleached sulfate eucalyptus (hardwood). Scale bar = 200 μm.*

**Table 2.1**  *The cell dimensions of typical hardwood and softwood.*

|                     | *Length* (mm) | *Width* (mm)  | *Aspect ratio* |
| ------------------- | ------------- | ------------- | -------------- |
| Softwood tracheids  | 2–4           | 0.02–0.04     | 50–200         |
| Hardwood tracheids  | 1.1–1.2       | 0.014–0.04    | 28–86          |

growing season) and summer wood (cells synthesised in the later part of the season). Usually spring wood tracheids have thinner walls and larger diameters than summer wood tracheids. In addition to tracheids, there is a small number (less than 10%), of ray cells.

These are narrow short cells (usually less than 0.2 mm in length) which are often lost in screening of chemical pulps, but are usually found in mechanical pulps. In addition, there are also epithelial cells which, in the original wood, are found surrounding the spaces which make up the resin canals. These canals may be up to 0.3 mm in diameter, and contain large quantities of resin, the chemistry of which is discussed more fully later in this chapter.

In hardwoods, about 50% of the volume of the wood is made up of fibres and fibre tracheids, which are considerably shorter than softwood tracheids, being of the order of 0.5 to 3 mm, with an average of around 1 mm and with a very narrow width of around 20 $\mu$m. In addition to these cells, there are also vessel elements which are large empty cells and which vary considerably in size and shape. They are a series of broad, articulated cells (around 100 $\mu$m), which are very long (many centimetres) and their function is to channel sap in almost straight lines. In some species, they may account for up to 50 to 60% of the volumetric composition, but usually less than 10% by weight. Ray and parenchyma cells are also present in hardwoods, as they are in softwoods, although they are more abundant and exhibit a greater variety of form. The ray cells are thin-walled rectangular shaped cells which grow at right angles to the wood fibre. They may constitute from 5 to 35% of the volume of the original wood.

## Softwood Structure

Four distinguishable layers or groups of lamellae can be identified in mature softwood cells. These are the primary cell walls, and the three parts of the secondary cell wall (the outer, the middle and the inner secondary cell wall—sometimes referred to as S1, S2 and S3 layers). The primary cell wall is a thin membrane which surrounds the protoplast during cell division and subsequent enlargement. In the developing cell, it mostly consists of water but a substantial proportion of the solid material is cellulose in the form of microfibrils which are widely spaced and partially interwoven. These microfibrils must have the ability to move relative to one another as the cell enlarges. The microfibrils are believed to be deposited transversely but, as the cell elongates, the orientation becomes much less marked, due to longitudinal displacement. The secondary cell wall is formed within the primary wall and comprises a series of lamellae, which are much more ordered than in the primary cell. A schematic representation of the structures of the primary and secondary cell walls of a softwood tracheid is shown in Figure 2.3.

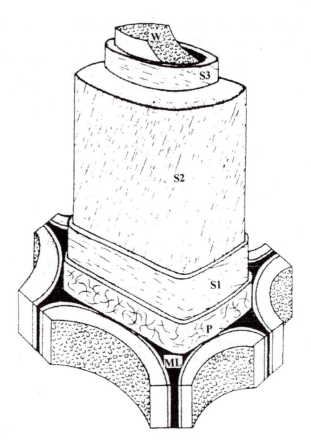

**Figure 2.3** *A schematic representation of the structure of the primary (P) and secondary (S1, S2 and S3) cell walls of a softwood tracheid (ML = middle lamella).*
(Source: Reproduced from 'Wood Ultrastructure', W.A. Cote Jr., University of Washington Press, Syracuse, NY, 1967).

The outer secondary cell wall (S1) is comparable in thickness to the primary wall and consists of four to six lamellae which spiral in opposite directions around the longitudinal axis of the tracheid. The main bulk of the secondary wall is contained in the middle secondary cell wall (S2), and may be as little as 1 $\mu$m thick in early woods and up to 5 $\mu$m in summer wood. The microfibrils of this part of the wall spiral steeply about the axial direction at an angle of around 10 to 20°. The inner secondary wall (S3), sometimes also known as the tertiary wall, is not always well developed, and is of no great technological importance.

The orientation of the microfibrils within the S2 layer has an important bearing on mechanical properties of the fibre such as its modulus of elasticity. In general, the smaller the angle that the microfibrils of the S2 layer make with the fibre axis, the greater is the stiffness of the fibre and the greater is its resistance to creep in response to axial stress. Figure 2.4 shows the relationship between the mechanical properties of single cotton fibres (elastic modulus and extension at break) and the mean fibrillar orientation of microfibrils within various layers of the cell wall as measured by the X-ray angle. The molecular architecture of the cellulose molecule in relationship to the microfibrils and the total cell wall is shown in Figure 2.5.

## Hardwood Structure

Although there are about twice as many hardwood as softwood trees throughout the world, hardwoods provide only around 25% of the world's wood pulp for paper making. This is because hardwood forests usually contain many different species and these have varying chemical requirements for pulping. The wood in the stem of hardwood trees is also usually a smaller proportion of the entire tree than in softwoods and, in addition, hardwood fibres are shorter and thicker walled and tend to have a higher hemicellulose content than softwood tracheids. This gives rise to a weaker fibre and therefore has an influence upon strength. Because of the presence of vessel elements there is also a relatively low yield of elongated cells.

## CHEMICAL COMPOSITION OF PAPER

The chemical composition of paper will depend greatly upon the chemical treatment which the wood has been subjected to during its conversion to pulp. When the pulp has received little or no chemical treatment, as in the case of pulp for newsprint, the chemical composition is very similar to that of the native wood. However, in those papers which have been chemically delignified, the composition may be very different. The natural compositions of native wood (softwoods and hardwoods) and the chemical pulps derived from them are shown in Table 2.2.

In general, these chemical treatments reduce the percentage of lignin, hemicellulose and extractives and increase that of cellulose. The chemistry of these processes is discussed more fully in Chapter

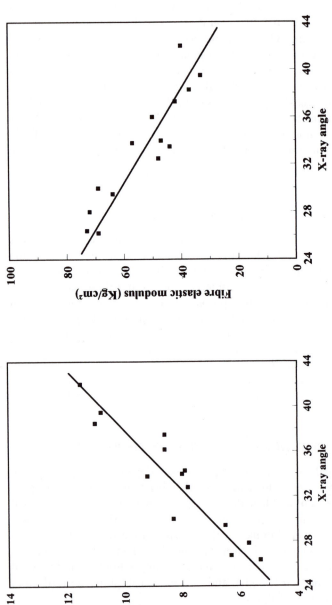

**Figure 2.4** *The effect of fibrillar angle upon the mechanical properties of the fibres when they are used to make paper.* (Source: Adapted from (*i*) 'Cell Wall Mechanics of Trecheids', M.R.E. London, Yale University, 1967, p. 169–170; (*ii*) 'A Microscopic Study of Coniferous Wood in Relation to its Strength Properties', H. Garland. *Ann. Missouri Botan. Gard.*, 1939, **26**, 1–95; (*iii*) 'Morphological Foundations of Fibre Properties', L.J. Rebenfeld, *J. Polymer Sci.*, 1965, **C9**, p. 91–112).

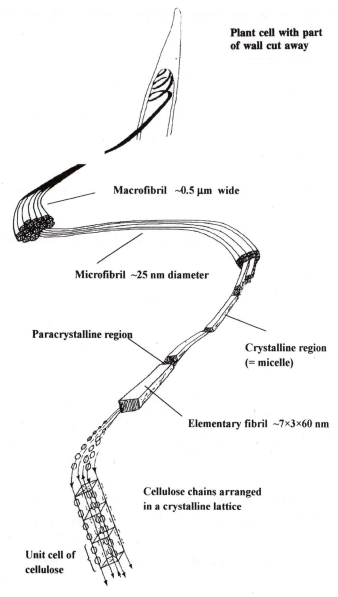

**Plant cell with part of wall cut away**

**Macrofibril** ~0.5 µm wide

**Microfibril** ~25 nm diameter

**Paracrystalline region**

**Crystalline region (= micelle)**

**Elementary fibril** ~7×3×60 nm

**Cellulose chains arranged in a crystalline lattice**

**Unit cell of cellulose**

**Figure 2.5** *The molecular architecture of the cellulose molecule showing its relationship to the microfibrils and to the total cell wall.*
(Source: Adapted from various sources including: P.A. Moss, PhD Thesis, University of Manchester, 1990; 'Electron Microscopy and Plant Ultrastructure', A.W. Robards, McGraw-Hill, NY, 1970).

**Table 2.2**  *The natural composition of native wood and the pulps derived from them.*
(Dated collected from various sources).

| Pulping process | Tree species | Cellulose (%) | | Hemicellulose (%) | | Lignin (%) | | Extractives (%) | |
|---|---|---|---|---|---|---|---|---|---|
| | | Wood | Pulp | Wood | Pulp | Wood | Pulp | Wood | Pulp |
| Sulfite | Spruce | 41 | 78.1 | 30 | 17.1 | 27 | 3.8 | 2 | 1.0 |
| | Birch | 40 | 81.6 | 37 | 12.2 | 20 | 4.1 | 3 | 2.1 |
| Kraft | Pine | 39 | 73.3 | 30 | 18.9 | 27 | 6.3 | 4 | 1.1 |
| | Birch | 40 | 63.6 | 37 | 31.8 | 20 | 3.7 | 3 | 0.9 |

3. For the moment, the chemistry of each of the individual groups of components will be considered.

### Cellulose

Cellulose is the primary structural component of the cell wall and, after removal of lignin and various other extractives, it is also the primary structural component of paper. Chemically, it is a semi-crystalline microfibrillar linear polysaccharide of $\beta$-1,4-linked D-glucopyranose (Figure 2.6).

Like most polysaccharides it is polydisperse with a high molecular weight. Its degree of polymerisation is typically between 10 000 and 15 000 glucose residues depending upon source and it is never found in a completely crystalline form, but occurs as a partly crystalline

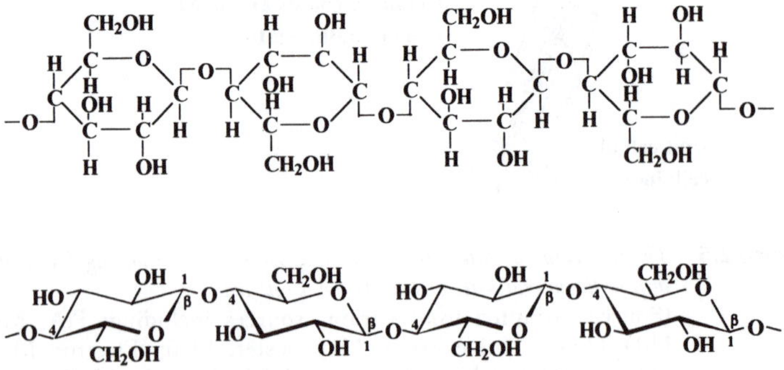

**Figure 2.6**  *The molecular structure of cellulose.*

and partly amorphous material. The degree of crystallinity is dependent upon the source of the cellulose. Cotton and various algal celluloses such as *Valonia*, are highly crystalline, whereas wood cellulose tends to be less so. Cellulose can also originate from bacterial sources, although these have no commercial use as fibre in paper manufacture, the best known example being *Acetobacter xylinum* which produces extra cellular cellulose as a small pellicle extending from its cell. It is not known why bacteria biosynthesise extra cellular cellulose, but it does not seem to have a structural function as it does in plants. Cellulose biogenesis in *Acetobacter* has been studied extensively and there are many parallels to its formation in plants where it is biosynthesised from uridine diphosphate-D-glucose (UDP-D-glucose) which is able to add one glucose unit to the growing polymer chain (Figure 2.7). The polysaccharide chains form crystalline domains after biosynthesis. The detailed molecular and crystal structure of cellulose is discussed more fully in Chapter 4.

## Hemicelluloses

The hemicelluloses are a group of non-structural, low molecular weight, mostly heterogeneous polysaccharides which are unrelated to cellulose and are formed biosynthetically by a separate route. They are not, as the name seems to imply, biosynthetic precursors of cellulose. Their function in the cell wall is poorly understood, but their molecular weight is too low for them to be major structural components (their degree of polymerisation is between 150 and 200). There has been speculation that they may have some function in water transport. They are most usually based upon polysaccharides of the hexoses D-glucopyranose, D-mannopyranose and D-galacto-pyranose and the pentoses D-xylopyranose and L-arabinofuranose. Smaller amounts of D-glucuronic and/or D-galacturonic acid and their 4-*O*-methylated derivatives are also usually present and the monosaccharide units are often partly acetylated. As can be seen from Table 2.2 a substantial amount of hemicellulose is retained in pulp even after its chemical delignification.

The principal hemicellulose present in softwoods is galactogluco-mannan which constitutes about 20% of the dry weight. This consists of a linear β-1,4-linked D-glucopyranose and D-mannopyran-ose backbone with α-1,6-linked D-galactopyranose residues as single side chain substituents. The galactose substituents may be of high or low frequency depending upon the source of the galactogluco-mannan. In the low galactose containing types, the ratio of galactose

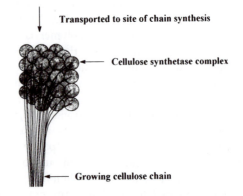

**CH₂OH**

**D-glucopyranose**

**α-D-glucopyranosyl-1-phosphate**     **Uridine triphosphate (UTP)**

**Uridine diphosphate-D-glucose (UDP-D-glucose)**

Transported to site of chain synthesis

Cellulose synthetase complex

Growing cellulose chain

**Figure 2.7**   *Biosynthesis of cellulose from D-glucopyranose in land plants.*
(Source: Adapted from various sources including: (*i*) 'Wood chemistry', E. Sjostrom, 2nd edition, 1993, p. 52; (*ii*) 'Cellulose Biosynthesis', D.F. Delmer, *Ann. Rev. Plant Physiol.*, 1987, **38**, 259–290; (*iii*) Biosynthesis in Plant Cell Walls', D.F. Delmer, in 'The Biochemistry of Plants', vol. 14, Academic Press, San Diego, 1988, pp. 373–420).

to glucose to mannose is about 0.1:1:4, whereas in the high galactose containing variety the ratio is 1:1:3. In addition, the polysaccharide is usually partially acetylated.

The second important group of hemicelluloses found in softwoods are the arabino-(4-$O$-methylglucurono)xylans, which may make up from 5 to 10% of the dry weight of the wood. These consist of $\beta$-1,4-linked D-xylopyranose units, partially substituted by L-arabino-furanose (at the 3 position) and by 4-$O$-methyl D-glucuronic acid (at the 2 position). The frequency of these substituent groups is around 1 to 2 residues per 10 xylose units.

Smaller amounts of other hemicellulose polysaccharides are also found in softwoods. In particular, larch contains an unusually large amount of arabinogalactan, which is usually only a minor components of other wood species.

The major group of hemicelluloses found in hardwoods are the glucuronoxylans. These consist of a $\beta$-1,4-linked D-xylopyranose backbone with 4-$O$-methyl D-glucuronic acid substituents linked $\alpha$-1,2. In addition, the 2,3 positions of the xylose backbone may be partially acetylated. The glucuronoxylan content of hardwood is typically between 15 and 30% by weight of the wood. Hardwoods also often contain small amounts of glucomannan, typically around 2 to 5% by weight. This is very similar to the galactoglucomannan found in softwoods, in that the backbone consists of $\beta$-D-glucopyranose and $\beta$-D-mannopyranose units 1,4-linked to each other. The proportion of glucose and mannose may vary from 1:2 to 1:1 depending on wood species.

The hemicelluloses are much more easily hydrolysed by acids than cellulose, but they tend to be more stable to alkali. It is widely recognised that they are beneficial to pulp and paper properties, although the reasons are not well understood. The tensile strength of paper, for example, generally correlates positively with the hemi-cellulose content. There is some evidence that they become adsorbed to fibre surfaces during pulping and mechanical refining where they might be expected to assist in inter-fibre bonding. They may also, because of their non-crystalline hydrophilic nature, contribute towards the swelling of the pulp and hence the conformability of the wet fibres during sheet formation.

The hemicelluloses are soluble in alkali, and can therefore be readily separated from the cellulose component by alkali extraction. However, this can only be done when the wood has first been delignified. This is because they are probably linked to lignin *via* covalent ester linkages (see Chapter 3) which need to be cleaved before the hemicelluloses can be solubilised. It might therefore be expected that they would be solubilised and removed from the wood during alkaline pulping. However, this is not the case. The glucuronoxylans for example are solubilised in the early stages of alkali pulping, but are then reprecipitated onto the fibre surfaces and presumably also back into the cell wall in the later stages of pulping. This is due partly to a decrease in their solubility as the alkali is consumed during the pulping process, and partly to structural modifications such as the removal of uronic acid groups which make the polysaccharide less soluble in alkali. Because of this, the hemicellulose content of alkaline pulps (Kraft) is higher than that of acidic sulfite pulps (Table 2.2). This reprecipitation may also be a contributory factor in making Kraft pulps stronger than sulfite pulps. However, it is not the only explanation and this point is discussed more fully in Chapter 3.

## Lignin

Lignin is an aromatic polymer, the structure of which is extremely complex and is discussed more fully in the following chapter. Almost all of its properties are undesirable for paper-making applications, and the highest qualities of paper are usually made from pulps from which most of the lignin has been removed. It causes paper to become brittle, and it is also oxidised photochemically to form coloured by-products which give rise to yellowing and discoloration. Newsprint is the best example of this, but all mechanical pulps in which the lignin is still largely present display this effect.

## Resins and Extractives

Wood contains a small proportion (usually less than 5%) of components which are extractable by organic solvents such as ethanol or dichloromethane. The proportion of these extractives varies in hardwoods and softwoods and also between species. Although many of these substances are removed during the chemical pulping process, some may still be retained in the final sheet of paper. Their chemical composition is very varied, and they include alkanes, fatty alcohols and acids (both saturated and unsaturated), glycerol esters, waxes, resin acids, terpene and phenolic components. The proportion which remains in pulp and paper depends upon the pulping process used. In general, acidic components such as the resin and fatty acids are relatively easily removed by alkali by conversion to their soluble carboxylate salt form but, in acidic pulping, they are not so readily solubilised.

Some useful by-products of pulping are derived from these extractives, the most important of which are turpentine and tall oil. Turpentine is a mixture of bicyclic hydrocarbons with the empirical formula $C_{10}H_{16}$, the dominant components of which are $\alpha$- and $\beta$-pinene (Figure 2.8). They are produced as volatile by-products at a yield of around 4–5 litres per tonne of wood (for pine) and are used as a solvent and as a chemical feedstock.

Tall oil is made up mostly of resin acids with around 10% of neutral components. These resin acids are isomers or structurally close relatives of abietic acid (Figure 2.9) and are used as anti-slip agents, as a chemical feedstock and as paper-sizing agents (see Chapter 7).

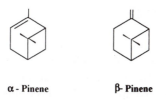

α - Pinene      β- Pinene

**Figure 2.8** *α- and β-pinene obtained as by-products from the pulping of wood.*

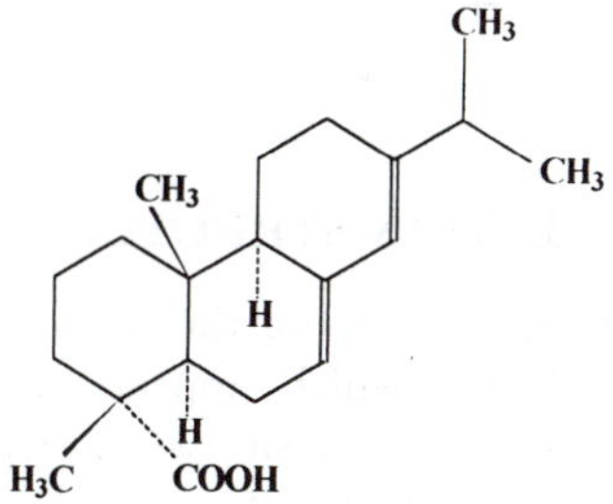

$CH_3$

$CH_3$

$CH_3$

$CH_3$

H

H

$H_3C$    COOH

**Figure 2.9** *The molecular structure of abietic acid—the dominant component of wood rosin.*

*Chapter 3*

# The Chemistry of Lignin and its Removal

## INTRODUCTION

In order to produce high quality paper with good strength properties it is necessary to remove the lignin from the wood or fibre matrix. Lignin, by virtue of photooxidation, discolours with age and causes the sheet to become brittle. The ideal pulping process would therefore completely dissolve it, whilst causing no loss or degradation to the carbohydrate component. However, ideal processes do not exist and all the current methods of pulping are a compromise. In order to understand the chemistry of lignin dissolution it is helpful to understand its molecular structure and something of its distribution within the cell wall.

## LIGNIN STRUCTURE

Lignin comprises about 17–33% of the dry weight of wood. It is a complex aromatic polymer which appears to function both as a strengthening agent in the composite wood structure and also as a component which assists in the resistance of the wood towards attack by micro-organisms and decay.

Whilst it is not possible to give a completely detailed structure for lignin, a great deal is known about the molecule. All lignins appear to be polymers of 4-hydroxycinnamyl alcohol (*p*-coumaryl alcohol) or its 3- and/or 3,5-methoxylated derivatives, respectively coniferyl and sinapyl alcohol (Figure 3.1).

The contribution of each of these three monomers to the lignin macromolecule differs depending on the source of the lignin. Gymnosperm (softwood) lignin is based only on coniferyl alcohol,

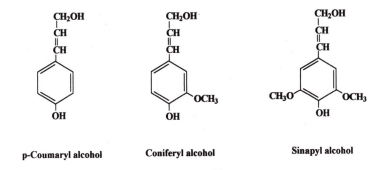

**Figure 3.1**   *The three monomer repeat units of lignin.*

whereas angiosperm (hardwood) lignin is a mixed polymer based on both coniferyl and sinapyl alcohols, and lignin obtained from grasses contains all three alcohols. Structural studies of lignin have proved to be extremely difficult and have been complicated by the fact that there are many bond types in the polymer. These bond types are either carbon–carbon or carbon–oxygen–carbon (ether) and they may involve both the aromatic rings and the three carbon atoms in the side chain. Figure 3.2 and Table 3.1 show the main linkage types in lignin and also the approximate frequency of occurrence in softwoods and hardwoods.

On the basis of analytical and degradative investigations, a partial structure has been proposed for a spruce lignin fragment of 16 aromatic units (Figure 3.3). The average number of methoxy groups is one per aromatic ring and, in addition, there are approximately 0.1 to 0.3 free phenolic groups per aromatic ring in the molecule.

## THE BIOSYNTHESIS AND BIOGENESIS OF LIGNIN IN PLANT CELL WALLS

### Lignification

The formation of lignin is unique to vascular plants and, in the case of wood, it provides the tree with unique strength and elastic properties. Primitive plants such as fungi and algae, which do not have differentiated cell tissues, do not contain lignin. Lignin is not uniformly distributed throughout wood, and the use of ultraviolet microscopy shows that it is concentrated in the inter-cellular spaces and is also present, but at a lower concentration, in the cell walls. The distribution of lignin in early wood tracheids of black spruce is shown in Figure 3.4.

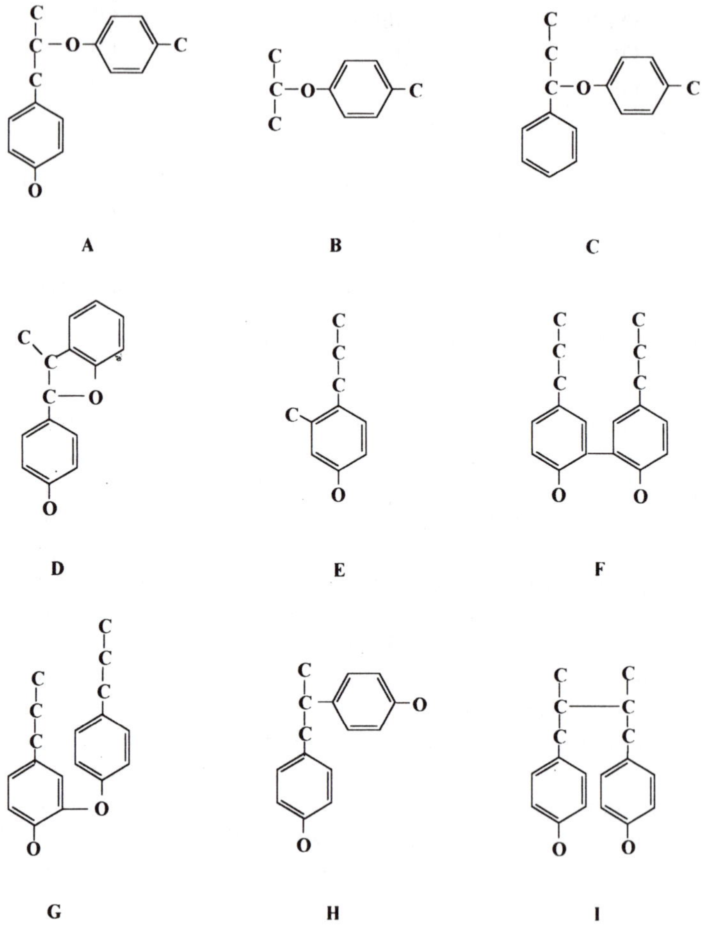

**Figure 3.2**   *The main linkage types in softwood and hardwood lignin.*
(Source: E. Alder, *Wood Science and Technology*, 1977, **11**,
169–218).

Although rich in lignin, the middle lamella, because of its relatively small volume, is not the region in which most of the lignin is
located. The secondary cell wall is lignified to a significant degree
and, because of its relatively large tissue volume in comparison to
the middle lamella, most of the lignin in wood is in fact located here
(Table 3.2). This fact has important consequences for pulping
chemistry, because pulping chemicals, if they are to attack the lignin
macromolecule effectively, must be able to penetrate the cell wall
and allow dissolution of the lignin.

**Table 3.1** *Frequency of occurrence of bond types A–I (see Figure 3.2) in softwood and hardwood lignin.*
(Source: E. Alder, *Wood Science and Technology* 1977, **11**, 169–218).

| Bond | Percentage (%) | |
|------|----------|----------|
| | Softwood | Hardwood |
| A | 48 | 60 |
| B | 2 | 2 |
| C | 6–8 | 6–8 |
| D | 9–12 | 6 |
| E | 2.5–3 | 1.5–2.5 |
| F | 9.5–11 | 4.5 |
| G | 3.5–4 | 6.5 |
| H | 7 | 7 |
| I | 2 | 3 |

## Biosynthesis

A clear understanding of lignin deposition in the cell wall is not yet possible, but a number of facts are known. Lignin precursors of the phenylglucoside type are formed either in the region of the cambium (the zone of new cell synthesis) or within the lignifying cell itself. Lignification is thus initiated in the differentiated wood cells from the primary walls adjacent to the cell corners and then extends into the inter-cellular area, the lamella, and thereafter to the primary and secondary cell walls.

The primary pathway to lignin biosynthesis is the formation of the two key amino acids: L-phenylalanine and L-tyrosine, both of which are formed from shikimic acid (Figure 3.5). In wood, lignin is synthesised from L-phenylalanine only, but the grasses do so from both L-phenylalanine and L-tyrosine. The 4-hydroxycinnamic acid which is formed from these amino acids can then be consecutively hydroxylated and methoxylated at the 3- and 5-positions to give ferrulic and sinapic acids which are then reduced to the corresponding alcohols — *p*-coumaryl alcohol, coniferyl alcohol and sinapyl alcohol (Figure 3.1). The next stage in the biosynthesis is the polymerisation, by dehydrogenation, of these three alcohols by either laccase/oxygen or peroxidase/hydrogen peroxide. These reactions produce phenoxide radicals in which the unpaired electron may be delocalised (Figure 3.6).

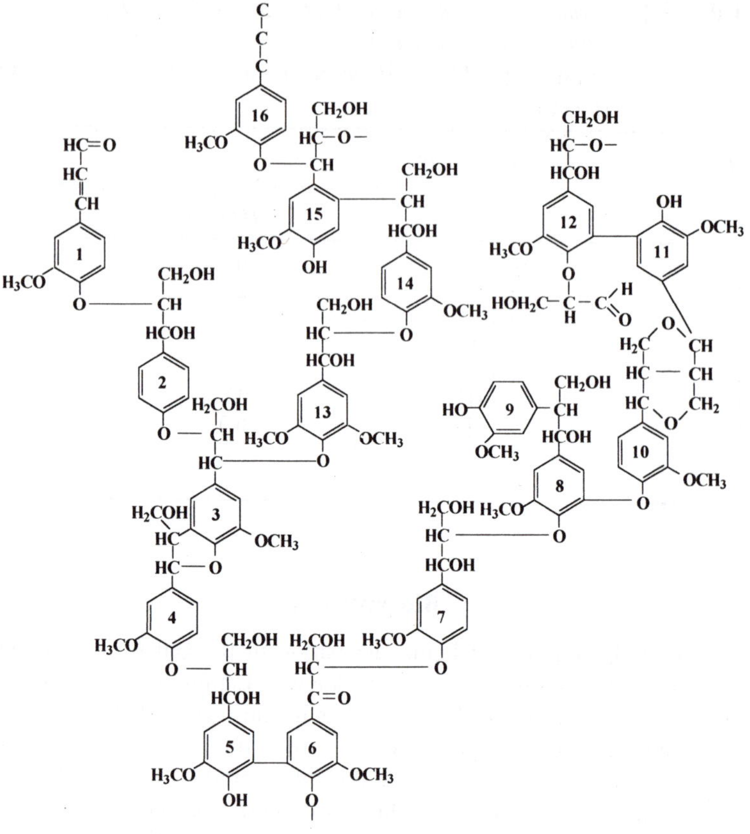

**Figure 3.3**   *Partial structure (16 aromatic units) of a spruce lignin fragment.*
(Source: E. Alder, *Wood Science and Technology*, 1977, **11**, 169–218).

The radicals thus formed may then couple non-enzymatically in an apparently random fashion to give dimers, trimers and higher oligomers (Figure 3.7). These oligomers and also the lignin produced by their further coupling are optically inactive, despite the presence of chiral centres in the side chain of the phenylpropane units. This makes lignin one of the most unusual of all natural products and it is atypical of many biopolymers such as nucleic acids, proteins and polysaccharides which are all chiral in nature.

It is possible to prepare a lignin-like polymer, known as dehydrogenated polymerisate (DHP) in the laboratory by treating coniferyl alcohol *in vitro* under aerobic conditions with a phenoloxidase

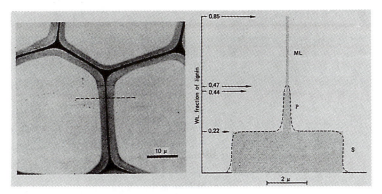

**Figure 3.4**   *Distribution of lignin in earlywood tracheids of black spruce.*
(Source: Reproduced from 'Pulp and Paper Chemistry and
Chemical Technology', ed. J.P. Casey, Wiley-Interscience, New
York, 1980, Vol. 1, p. 4. Adapted originally from: (i) B.J.
Fergus, A.R. Procter, J.A.N. Scott and D.A.I. Goring, *Wood
Science and Technology*, 1969, **3**, 117–138 and (ii) K.V. Sarkanen
and C.H. Ludwig (eds.) in, 'Lignins—Occurrence, Formation,
Structure and Reactions', Wiley-Interscience, New York, 1971).

**Table 3.2**   *Distribution of lignin in softwoods and hardwoods in the various
morphological zones.*
(Source: B.J. Fergus, A.R. Procter, J.A.N. Scott and
D.A.I. Goring, *Wood Science and Technology* 1969, **3**,
117–138).

| Wood | Morphological region | Tissue volume (%) | Lignin (% of total) | Lignin concentration (%) |
|------|----------------------|-------------------|---------------------|--------------------------|
| Softwoods | Secondary wall | 91 | 77 | 23 |
| | Compound middle lamella | 7 | 13 | 55 |
| | Cell corner | 3 | 11 | 93 |
| Hardwoods | Secondary wall | 73 | 60 | 19 |
| | Compound middle lamella | 5 | 9 | 40 |
| | Cell corner | 2 | 9 | 85 |

enzyme. The DHP thus formed appears to be closely related to
natural lignins in terms of its functional group content, and its UV,
IR and $^1$H and $^{13}$C NMR spectra. DHP has been used as a model
substrate for studies in lignin biodegradation, and it is hoped that

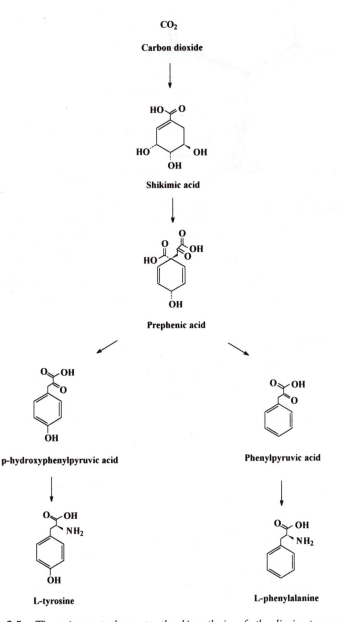

**Figure 3.5** *The primary pathway to the biosynthesis of the lignin precursors L-tyrosine and L-phenylalanine.*
(Source: H. Higuchi, M. Shimada, F. Nakatsubo and M. Tanahashi, *Wood Science and Technology*, 1977, **11**, 155).

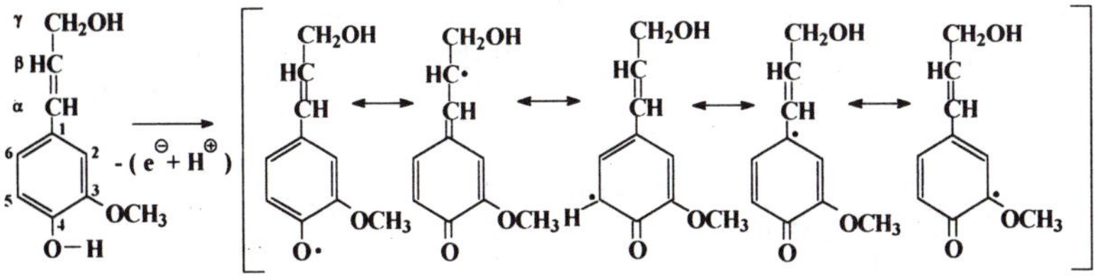

**Figure 3.6** *Resonance forms of phenoxide radicals generated in biosynthesis.*
(Source: Reproduced from 'Lignin Biodegradation: Microbiology, Chemistry and Potential Applications', eds., T.K. Kirk *et al*, CRC Press, Florida, 1980, Vol. 1, p. 6).

such studies will in due course lead to the development of commercial biochemical and biotechnological approaches to lignin dissolution from wood. However, at the moment, this seems some years away.

If the five resonance forms of the phenoxy radical (Figure 3.6) can couple to any other phenoxy radical, the theoretical number of dimeric structures possible is 25. The relative frequency of involvement of individual sites in the phenolic coupling reaction depends on their relative electron densities. Quantum mechanical calculations predict that the high electron densities at the phenolic oxygen atom and the $\beta$ carbon atom would give rise to a high proportion of $\beta$-$O$-4 linkages, which is indeed observed to be the case (Table 3.1).

These lignin precursors then continue to polymerise by a similar mechanism leading to a three-dimensional branched network polymer. The size of this polymer is one of the most difficult problems to resolve in lignin chemistry. The largest computer-simulated model which is in keeping with the all experimental observations of lignin biosynthesis is based on 81 phenylpropane units and has a total molecular weight of about 15 000. However, molecular weights of isolated lignins have been determined within the range 2000 to over 1 million. Although a great deal of work has been done in attempting to determine the molecular distribution of lignin within wood, the resolution of this problem has been difficult and it has even been suggested that lignin may exist as one single molecule in its native environment. Since it is formed by enzymatic dehydrogenation, followed by random coupling reactions, it is possible because of the presence of free phenolic hydroxyl groups that its structure never ceases to grow. However, this fact although important in many ways may not be too important from the perspective of the pulping

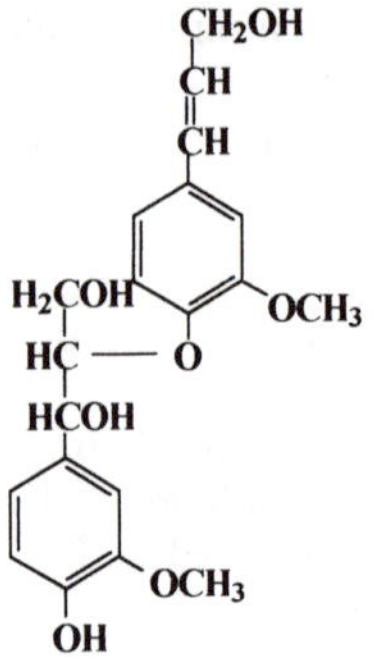

**Figure 3.7** *Formation of oligolignols by non-enzymatic coupling of phenoxide radicals.*
(Source: 'Lignin Biodegradation: Microbiology, Chemistry and Potential Applications', eds., T.K. Kirk *et al*, CRC Press, Florida, 1980, Vol. 1, p. 6).

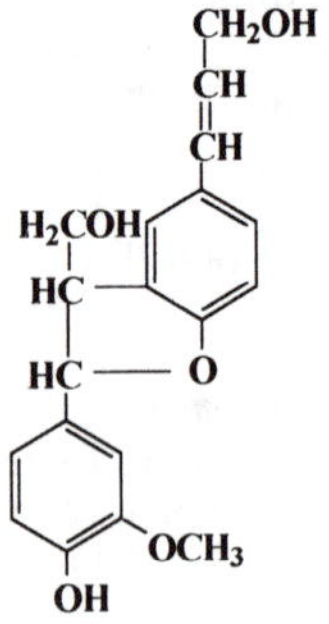

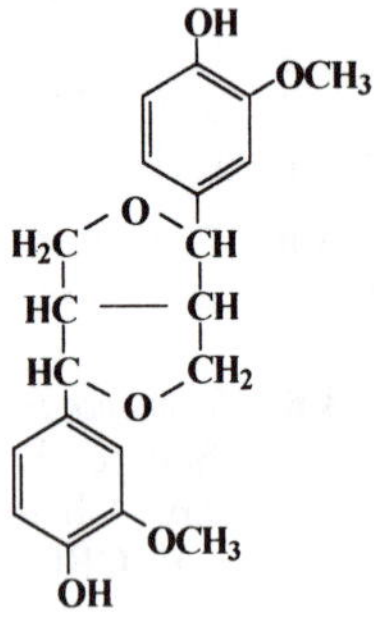

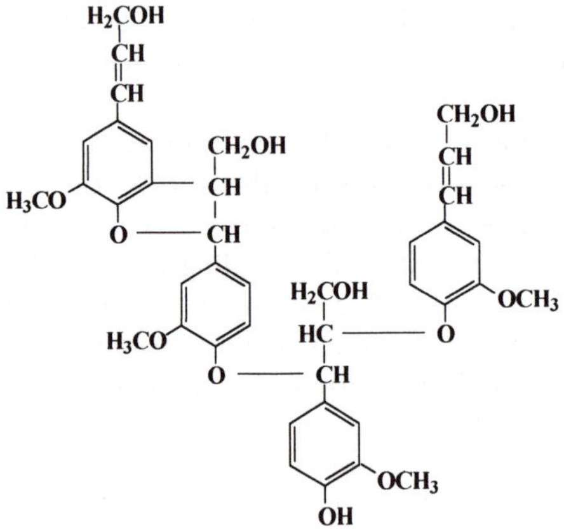

chemist, whose main aim is to degrade the lignin structure in order to achieve dissolution.

## The Association of Lignin and Carbohydrates

Lignin and carbohydrates exist in close association in the wood structure and there is now strong evidence to suggest that formal covalent links exist between the lignin macromolecule and carbohydrate components of the wood structure. More probably, lignin is bound by an ester linkage to 4-O-methyl-D-glucuronic acid residues which decorate the xylan backbone [Figure 3.8(a)]. Perhaps the most convincing evidence for this is the demonstration that guaiacyl-glycerol-β-guaiacyl ether reacts at the C-α position with the carboxy group of tetra-O-acetyl-β-D-glucuronic acid to give an ester linkage. The NMR spectrum of this product is shown in Figure 3.8(b).

## DISSOLUTION OF LIGNIN DURING PULPING

The mechanical and chemi-mechanical pulping processes remove very little lignin and these will be discussed only briefly here. The most widely used commercial methods of more extensive lignin removal are based on aqueous, high temperature extraction procedures at acidic, neutral or alkaline pH. A discussion of the chemistry of these processes follows. Large-scale delignification is not currently carried out by the use of organic solvents, although this is perfectly possible and well-suited to take advantage of the structural differences between the carbohydrates and lignin. The main obstacle to their commercial use is the recovery of very large quantities of organic solvents. However, in recent years pilot-scale operations of several thousand tonnes per year of pulp produced by organic solvent processes have been successfully employed.

### Mechanical Pulping

In its simplest form, mechanical pulping involves the conversion of raw wood into paper-making pulp by use of mechanical means only and does not involve the removal of lignin. There are many

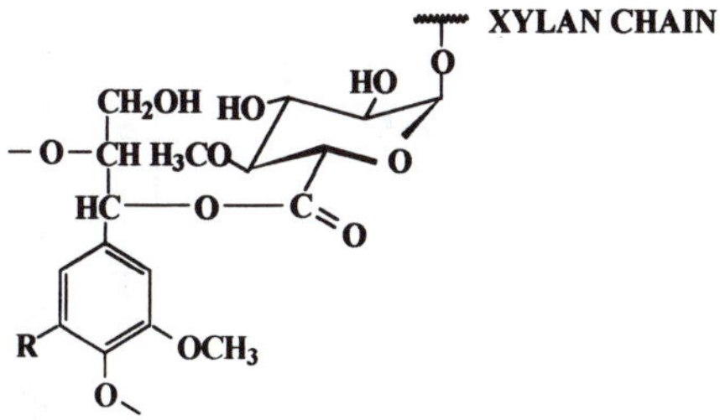

**Figure 3.8(a)** *Possible ester linkage between an α-OH of lignin and a carboxyl group of a 4-O-methyl-β-D-glucuronic acid residue on the xylan backbone.*

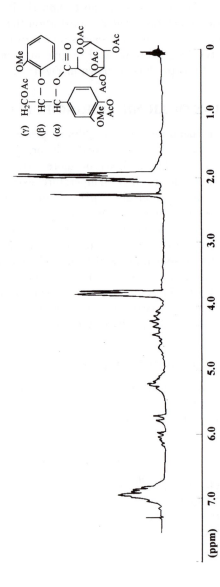

**Figure 3.8(b)**  *The $^1H$ NMR spectrum of the ester formed between guaiacylglycerol-β-guaiacyl and tetra-O-acetyl-β-D-glucuronic acid.* (Source: K. Tanaka, F. Nalatsubo and T. Higuchi, *Moluzai, Gakkaishi,* 1976, **22**, 58).

variations of the basic method which, in some cases, include mild chemical pretreatments. Until 1960, virtually all mechanical pulps were produced by what was known as the stone-ground wood process (SGW). In this process, blocks of wood are pressed against an abrasive rotating stone surface in an orientation parallel to the axis of the stone, so that the grinding process does as little damage as possible to the fibres. A typical grinder arrangement is shown in Figure 3.9.

Since 1960, mechanical pulps have been increasingly produced by refining processes in which the wood is introduced into a disk refiner *via* a screw feeder. As the wood moves into the refining zone, it is progressively broken down into smaller fragments and finally into fibres. Water is supplied to the refiner to control the consistency, and in some cases, chemicals are also added. Sometimes, the wood is steamed under pressure for a short period before the refining process. This softens the chips and produces pulp with a higher percentage of long fibres, and which is therefore stronger. This process is known as thermo-mechanical pulping and is now one of the major processes for the production of newsprint. The addition of

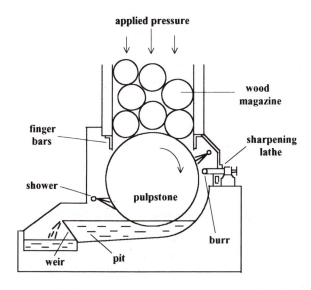

**Figure 3.9** *A typical grinder arrangement for the stone-ground wood (SGW) process.* (Source: Adapted from 'Handbook for Pulp and Paper Technologists', G.A. Smook, Angus Wilde Publications, Vancouver, 1992, p. 47).

chemicals, either prior to or during the refining process can significantly improve the quality of the resulting pulp. Sulfonation, using sodium sulfite, at typically between 1 and 5% based on the dry weight of the wood, is often used. The lignin, although not significantly removed, becomes partially sulfonated which has the effect of softening the wood in a permanent way. This sulfonation process may also be carried out at an intermediate stage in the pulping process, when the pulp has a larger surface area, and this produces a pulp with improved printing qualities and with improved wet web strength.

## Chemical Pulping: General Principles of Aqueous Lignin Dissolution

In the chemical pulping processes lignin is dissolved from wood at high pressures and temperatures under aqueous alkaline, neutral or acidic conditions. However, these conditions are severe and also cause degradation of the carbohydrate component by both lowering its molecular weight, hence reducing its strength, and by causing it to be partially solubilised, thus reducing the pulp yield. The aim of an ideal aqueous lignin extraction procedure would be to render the lignin soluble without causing any degradation to the carbohydrate components. This is not at the present time achievable, and all processes attempt to minimise the amount of carbohydrate degradation and maximise the amount of lignin dissolution (Figure 3.10).

Lignin is water-insoluble and it contains only a few hydrophilic functionalities (primarily the phenolic groups). In order to increase its solubility, aqueous pulping processes therefore seek either to introduce water solubilising groups, or to reduce the degree of polymerisation or both.

## Lignin Dissolution in Acidic Systems

Although acidic pulping methods have largely been displaced over the past 50 years by neutral and alkaline processes, there is still a significant amount carried out. Acid sulfite pulping uses combinations of sulfur dioxide and water at high temperatures and pressures. An appropriate base is used to control the pH and, although usually acidic, it is possible to perform these reactions at neutral or even alkaline pH. The most active nucleophile present is the bisulfite ion,

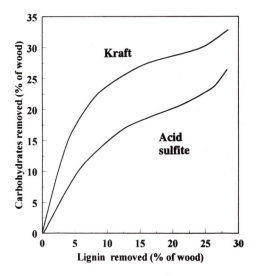

**Figure 3.10** *Relative amounts of lignin and carbohydrate removed during the Kraft and acid sulfite processes.*
(Source: Adapted from E. Sjostrom, 'Wood Chemistry', Academic Press, London, 1992, p. 124).

which arises *via* the following equilibria when sulfur dioxide is dissolved in water:

$$SO_2 + H_2O \rightleftharpoons H_2SO_3$$

$$H_2SO_3 \rightleftharpoons H^+ + HSO_3^-$$

$$HSO_3^- \rightleftharpoons H^+ + SO_3^{2-}$$

The concentration of sulfur dioxide, bisulfite ion and sulfite ion will be a function of pH, and the mole percent of the bisulfite ion is a maximum at around pH 4 (Figure 3.11).

At lower pH, sulfur dioxide is mostly present and, as the pH increases from 4 to 9, the proportion of sulfite ion increases and that of the bisulfite ion decreases. Usually either calcium or magnesium bases are used to control the pH within the acidic range. The pH is dependent upon the relative solubility of the calcium or magnesium sulfite, and upon the excess of sulfur dioxide which is used. Because of the greater solubility of magnesium sulfite, the magnesium-based

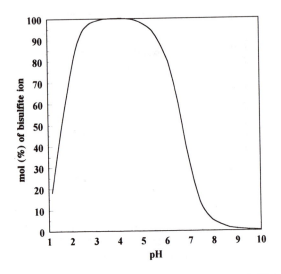

**Figure 3.11** *The mole percent of the bisulfite ion as a function of pH for an aqueous solution of sulfur dioxide.*
(Source: E. Sjostrom, P. Haglund, and J. Janson, *Svensk Papperstidning*, 1962, **65**, 855–869).

process can operate at a pH of around 4 to 5, whereas the calcium-based process operates at a lower pH.

At these pH values in both the calcium- and magnesium-based processes, the dominant species is the bisulfite ion, and it is this which is the primary nucleophile involved in delignification. The chemical treatment begins with an impregnation stage, essential for satisfactory delignification, in which the chips are submerged in the cooking liquor. Some degradation of the carbohydrate component occurs and the hemicelluloses are particularly vulnerable to acid hydrolysis, and are easily lost during acidic pulping processes. Cellulose, on the other hand, although partially depolymerised *via* acid hydrolysis of glycosidic linkages, is not dissolved to any great extent. This is because of its relative insolubility even at a low degree of polymerisation (DP). Cellulose oligomers are significantly soluble only at a DP below 6. Depolymerisation does lead to some strength loss, and this point is discussed more fully later in this chapter. Delignification produces soluble lignin sulfonic acids which retain a high degree of polymerisation. The rate of diffusion of the active chemicals into the reaction zone as well as the transport of products into solution are important factors in influencing the rate of delignification and, because of its high molecular weight, the molecule has difficulty in diffusing through the cell wall of the fibres and into solution, and this limits the extent of delignification. Lignin is solubilised not only by sulfonation but also to a lesser extent by hydrolysis. Sulfonation allows the introduction of hydrophilic sulfonic acid groups, whilst hydrolysis assists in cleaving ether linkages and thereby reducing molecular weight and creating new free phenolic groups. Both of these reactions increase the solubility of the lignin and their relative rates are a function of pH. Two important categories of hydrolysis and sulfonation reactions are shown in Figures 3.12 and 3.13.

The sulfonation reaction frequently involves the displacement of a

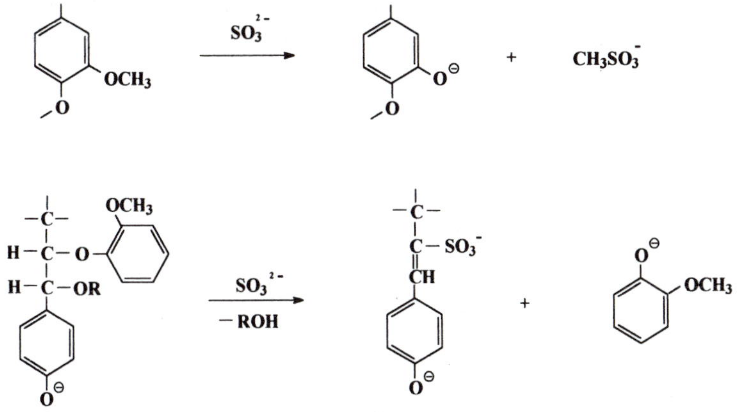

**Figure 3.12**  *Cleavage of ether linkages during sulfite pulping.*
(Source: Adapted from E. Sjostrom, 'Wood Chemistry', Academic Press, London, 1992, p. 112).

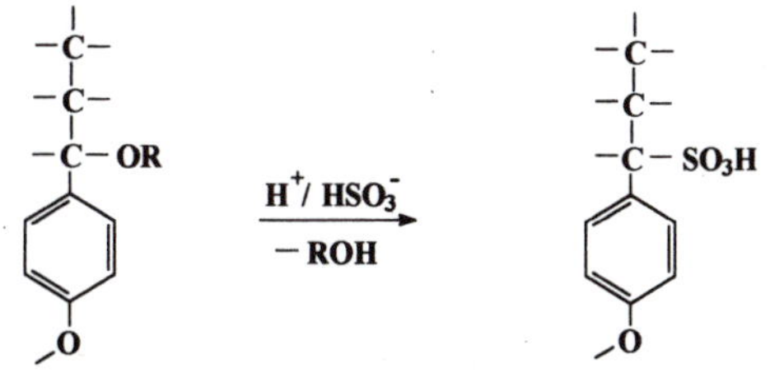

**Figure 3.13**  *Sulfonation of lignin α-carbon atoms during sulfite pulping.*
(Source: Adapted from E. Sjostrom, 'Wood Chemistry', Academic Press, London, 1992, p. 112).

hydroxy or alkoxy group from the $\alpha$-carbon atom of the phenylpropane side chain to form a carbonium ion which may then react with
a bisulfite ion to introduce a solubilising group into the lignin
molecule. However, other nucleophilic groups such as benzylium
ions are present in the cooking liquor and it is possible for
condensation reactions to take place with these which may lead to
an increase in molecular weight and a decrease in solubility (Figure
3.14).

## Lignin Dissolution in Aqueous Alkaline Systems

Alkaline delignification in the form of the Kraft or Sulfate process is
now the most widely used method of lignin removal. It uses a
mixture of sodium hydroxide and sodium sulfide—the latter being
produced in the recovery process by the reduction of sodium sulfate

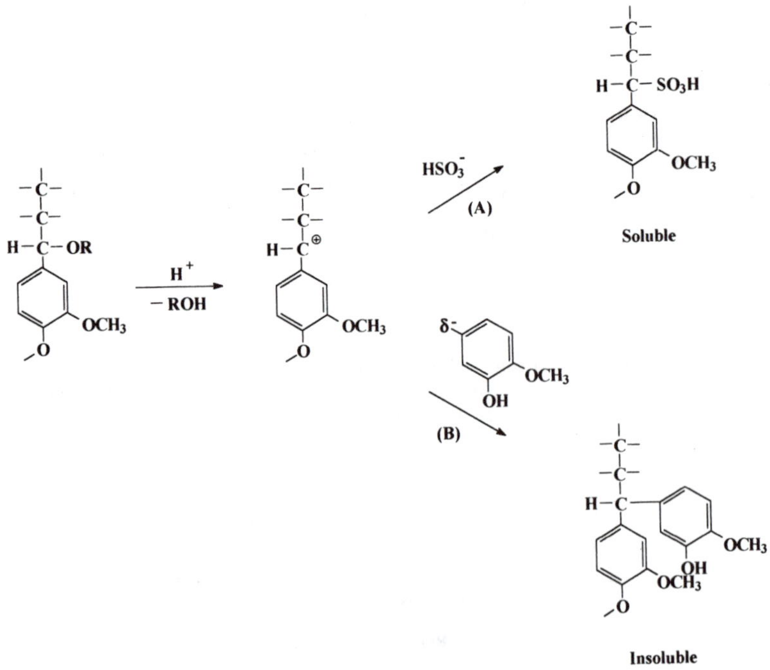

**Figure 3.14**   *Mechanism of nucleophilic substitution reactions of lignin during sulfite*
*pulping.*
(Source: Adapted from 'Pulp and Paper Chemistry and
Chemical Technology', ed. J.P. Casey, Wiley-Interscience,
New York, 1980, Vol. 1, p. 67).

(hence the name Sulfate process). The presence of sodium sulfide improves the efficiency of the process but it is not essential and, when it is absent, the process is simply referred to as alkaline pulping. The chemistry is substantially more complicated than that of the acidic sulfite process. In contrast to sulfite pulping, where the site of attack is usually the α-carbon atom of the side chain of the phenylpropane unit, in alkaline pulping the primary site of attack is the phenolic hydroxy group. The phenoxide ion which is generated is then able to eliminate an alkoxy group from the α-carbon atom and undergo nucleophilic attack at the same carbon atom by an OH⁻ or SH⁻ ion. This is followed by formation of an epoxide ring and cleavage of the β-aryl ether linkage (Figure 3.15).

The effect of this reaction is to depolymerise the lignin molecule and to generate free phenolic groups, both of which assist in its solubilisation. There is also some cleavage of methoxy groups on the aromatic rings which also leads to the formation of more solubilising phenolic groups.

The chemistry of this process is considerably more complex than described here and involves many other reactions including some

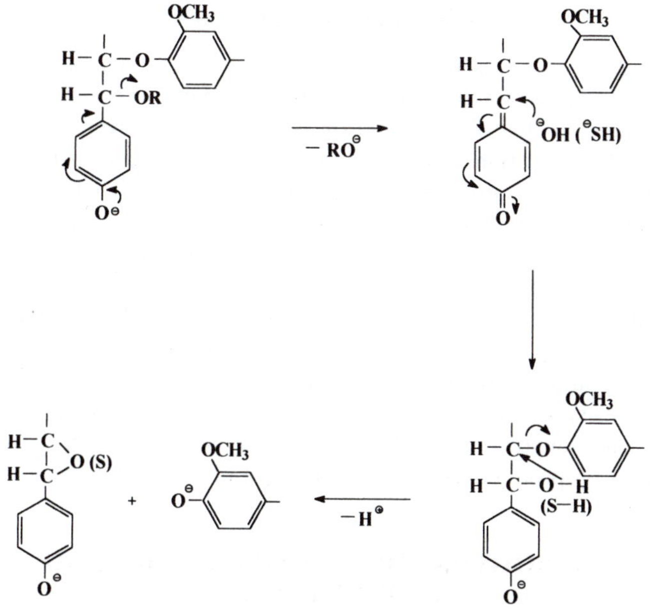

**Figure 3.15** *Mechanism of aryl ether cleavage during alkaline pulping.*
(Source: Adapted from E. Sjostrom, 'Wood Chemistry', Academic Press, London, 1992, p. 146).

repolymerisation *via* condensation reactions. However, the types of depolymerisation processes outlined in Figure 3.15 are probably very important and almost certainly make a major contribution to the dissolution process.

Recovery of inorganic chemicals is crucial to the cost effectiveness of the Kraft process. The black liquor which is obtained from delignification is rich in solubilised lignin and carbohydrate degradation products and, after concentration, is combusted in a recovery furnace. The carbon dioxide which is produced during combustion converts unused sodium hydroxide into sodium carbonate. In addition, the sodium sulfate is converted, under the reducing atmosphere of the furnace, to sodium sulfide.

The inorganic ash from the recovery furnace is then dissolved in water, and calcium hydroxide is added to precipitate out calcium carbonate and to convert the sodium carbonate to sodium hydroxide for reuse. The calcium carbonate is then separated out by sedimentation and is combusted to give calcium oxide which provides the calcium hydroxide which is used in the precipitation process.

## Solvent Pulping Processes

Although at the present time there are no commercial processes for the removal of lignin using organic solvents, this approach to delignification has been extensively researched, even as far as pilot-scale operations. The attraction of these processes is the avoidance of the production of an aqueous lignin-rich black liquor from which inorganics must be recovered, and also the advantages of relatively low energy solvent recovery. The basic principle is to dissolve the lignin in an organic solvent and recover the volatile solvents for further processing. A number of methods have been tried, the most commercially viable of which has been the use of methanol at high temperatures and pressures. This process allows the recovery of cellulose pulp after dissolution of the lignin in a mostly non-depolymerised form. Furfural derivatives are also produced as by-products from the breakdown of hemicelluloses.

## CARBOHYDRATE DEGRADATION DURING DELIGNIFICATION

The effect of chemical delignification on the carbohydrate fraction is predominantly that of pH, and it is important therefore to consider separately degradation in alkali and in acidic pulping systems.

## Carbohydrate Degradation during Alkali Pulping

When wood is subjected to the high temperature and pH of alkaline pulping processes, the carbohydrate components (celluloses and hemicelluloses) undergo various changes. Some is dissolved in the cooking liquors, particularly the readily alkali-soluble hemicelluloses, and some is degraded to form lower molecular weight products which may either remain in an insoluble form within the fibre matrix or be dissolved into the cooking liquors. The hemicelluloses have a relatively low DP to begin with (150–200) but this may be further reduced to about 40 during alkaline pulping. Many are also partially acetylated in their natural state, but these acetyl groups are very quickly cleaved under alkaline conditions. However, the most important degradative process in alkali pulping is that which is known as peeling, in which single monosaccharide units are sequentially removed from the reducing end of the chain. Both cellulose and hemicelluloses are susceptible to these reactions, and the pathway for cellulose is shown in Figure 3.16.

The reducing end of the polysaccharide isomerises to the keto-form, and this is followed by cleavage of the glycosidic linkage, thus leaving the chain shorter by one glucose residue but still with an exposed reducing end group. The product is soluble and undergoes further degradation to form a variety of soluble acids which exist in the cooking liquors as their carboxylate salts. The DP of the carbohydrates is reduced but not severely because the cleavage of glycosidic linkages is a sequential not a random process. On the other hand, these products of degradation, being of low molecular weight, are soluble and give rise to a loss of yield from the wood.

The extent to which peeling occurs is controlled by a much slower reaction, known as the stopping reaction. This also involves the reducing end of the molecule but in either an isomerised or a non-isomerised form (Figure 3.17).

The end group which is produced contains a carboxylic acid functionality which has an influence on the anionicity of pulp fibres (Chapter 6) but, in this form, it is resistant to further alkaline degradation. The hemicelluloses are also able to undergo the same type of peeling reaction but at different rates from each other and from cellulose. The $\beta$-1,4-xylans, for example, are more stable to alkaline degradation than the glucomannans.

Some cleavage of internal glycosidic linkages of cellulose does occur in alkaline pulping but only when temperatures are fairly high, and these cleavages are probably due to alkaline hydrolysis reactions involving the formation of 1,2-epoxides (Figure 3.18).

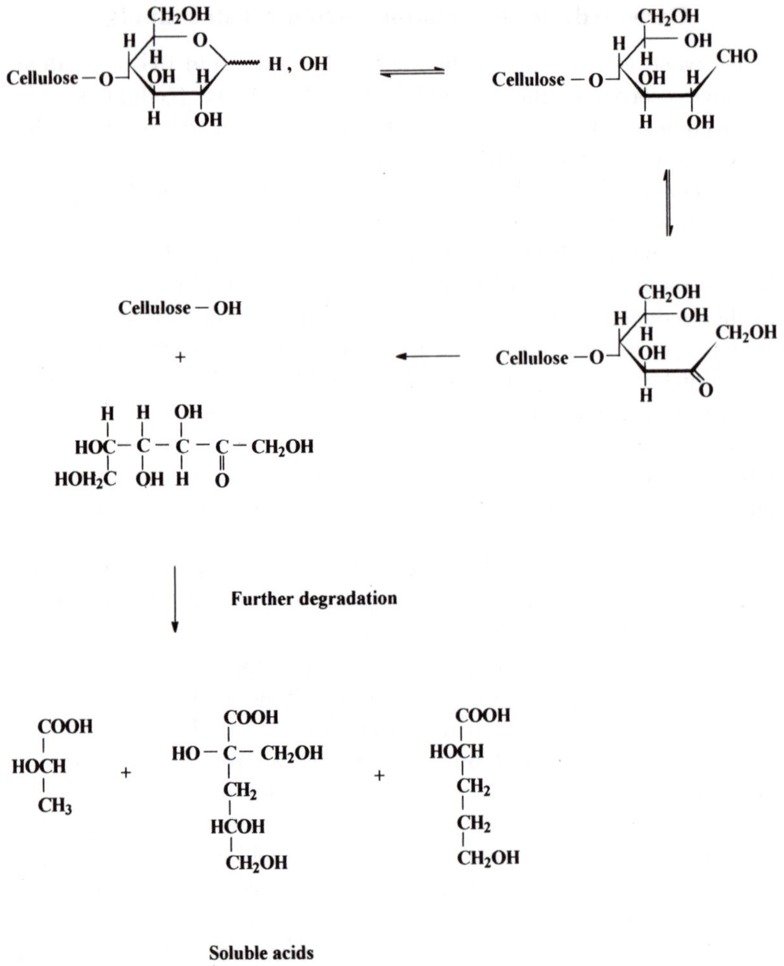

**Figure 3.16** *Depolymerisation of cellulose by alkaline peeling reactions.*
(Source: Adapted from E. Sjostrom, 'Wood Chemistry', Academic Press, London, 1992, p. 151).


## Carbohydrate Degradation during Acidic Pulping

In contrast to alkaline pulping, the acetyl groups of the hemicelluloses are relatively stable at low pH, as are the glucuronic acid residues of the xylans. The main effect of high temperature and low pH is the hydrolysis of the glycosidic linkages in the polysaccharide

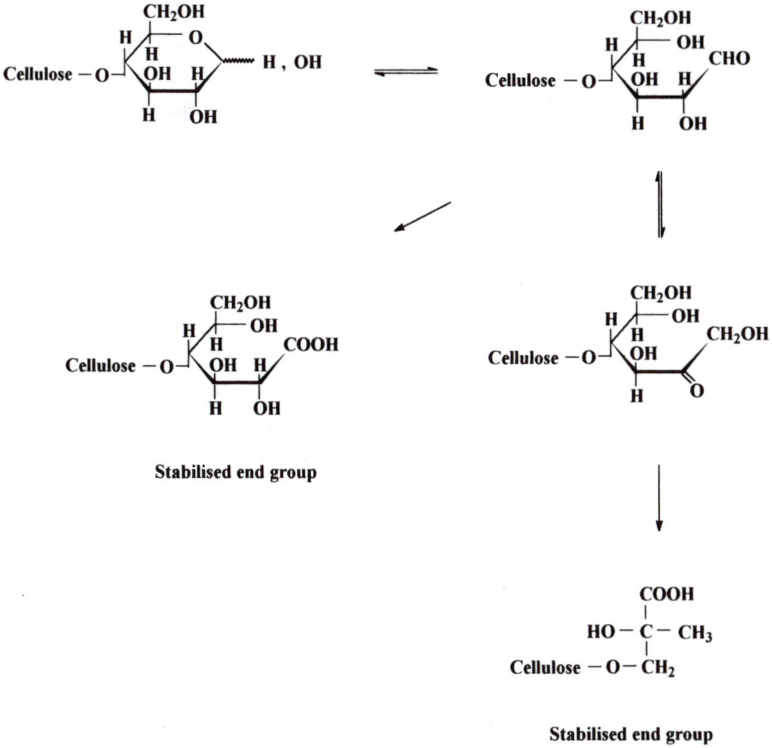

**Figure 3.17** *Alkaline chain stabilisation (stopping) reactions of cellulose.*
(Source: Adapted from E. Sjostrom, 'Wood Chemistry', Academic Press, London, 1992, p. 151).

chain. These processes seem not to be selective to the chain end but occur relatively randomly along the polysaccharide backbone. Because cellulose is extremely insoluble at degrees of polymerisation above 6, a substantial amount of depolymerisation may take place without very much solubilisation. However, the depolymerised pulp is significantly weakened and this probably accounts for the difference in strength between pulps from the acid and alkaline processes. The chemistry of the hydrolysis reaction is show in Figure 3.19. Hemicelluloses are particularly vulnerable to acid hydrolysis and undergo substantial depolymerisation and dissolution, which probably accounts for the lower levels of hemicelluloses in these pulps (Table 2.2, Chapter 2).

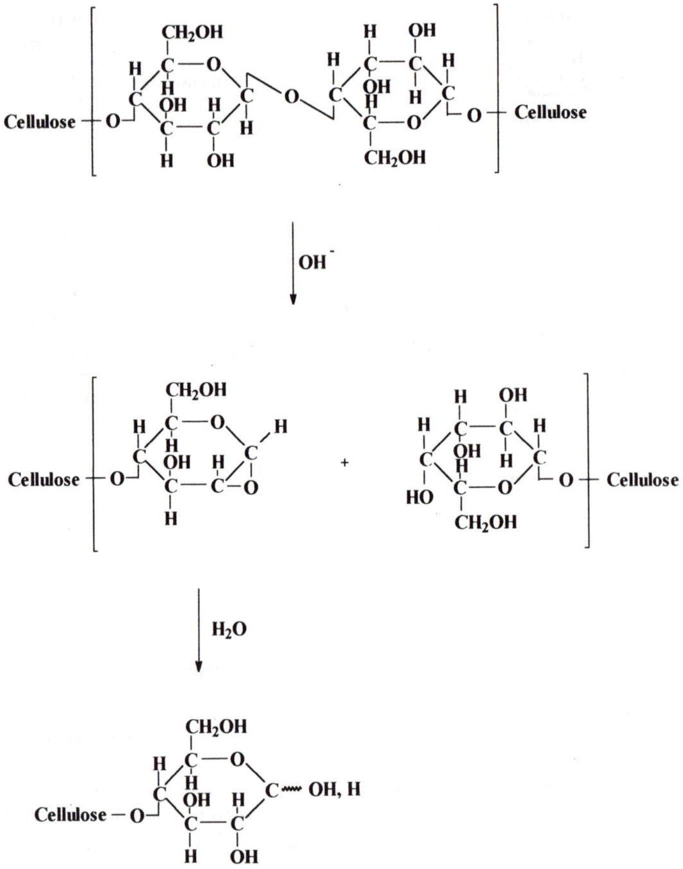

**Figure 3.18**   *Alkaline hydrolysis reactions of cellulose via 1,2-epoxides.*
(Source: Adapted from 'Pulp and Paper Chemistry and Chemical Technology', ed. J.P. Casey, Wiley-Interscience, New York, 1980, Vol. 1, p. 141).

## BLEACHING

The pulps obtained from chemical pulping are brown in colour and, whilst suitable for many applications, are unsuitable for printing and writing papers which require a bright white pulp. Both chemical and mechanical pulps often require bleaching for many end uses. The colour of these pulps is mainly due to residual lignin (~3–6%, see Table 2.2 in Chapter 2) although products derived from carbohydrate may also make a contribution. The approach to removal of colour is different in each case.

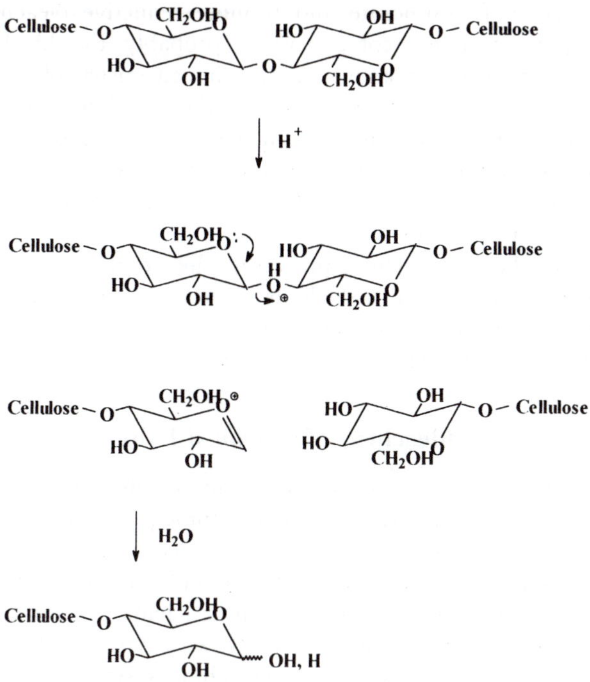

**Figure 3.19** *The hydrolysis of carbohydrate during acid pulping.*

The chemical reactions which are involved in the natural discoloration of wood during storage are extremely complex and poorly understood but are probably very similar to those involved in the discoloration of lignin-containing pulp and paper. Mechanical pulps retain a high proportion of the lignin in the original wood, and therefore have a much greater tendency towards discoloration than chemical pulps. However, the brightness of these pulps immediately after delignification is generally much lower than that of the wood from which they were produced because of the large increase in light absorption of the remaining lignin.

## Bleaching of Mechanical Pulp

There are two approaches to the bleaching of mechanical pulps. They may be either reductive or oxidative in nature. The reductive bleaching agents are usually bisulfite, dithionite or borohydride, and the oxidising agents are normally peroxide, hypochlorite, peracetic acid or ozone. In dithionite and bisulfite reductive bleaching, the most important active component is probably the bisulfite ion. Although very little is known about the chemistry of the colour removal process, it seems probable that the addition of bisulfite ion to carbonyl groups is one reason for colour removal and the reduction of o-quinones and coniferaldehyde groups is also possible. Quinoid structures are easily reduced by dithionite, but condensed quinones react more slowly.

In the oxidative bleaching processes, the decoloration of p- and o-quinones and of coniferaldehyde structures also seems to be involved. In the case of coniferaldehyde, the removal of the conjugated side chain is probably involved (Figure 3.20).

### Bleaching of Chemical Pulp

The bleaching of chemical pulps mainly involves the removal of residual lignin, and the lignin content of these pulps therefore gives a fairly good indication of the amount of bleaching chemical which will be required. Until recently, chlorine and compounds of chlorine have been the most widely used bleaching agents for chemical pulps but this situation is changing rapidly as environmental pressure builds for the use of non-chlorine bleaching systems (for a further discussion see Chapter 10). The most common approach is to use either chlorine in aqueous solution or chlorine dioxide, in combination with alkaline extraction stages. Various combinations of sequences are used to achieve different levels of brightness for different pulps.

Chlorination of pulp is usually carried out at fairly low consistency and its function is to convert the residual insoluble lignin in the pulp to compounds which are water or alkali soluble. Chlorine reacts very rapidly with pulp and most of it is consumed within a few minutes.

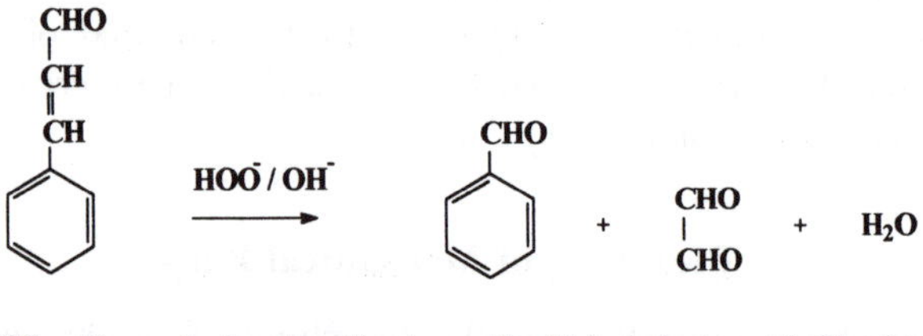

**Figure 3.20**  *Removal of conjugation in coniferaldehyde by peroxide bleaching.*
(Source: Adapted from 'Pulp and Paper Chemistry and Chemical Technology', ed. J.P. Casey, Wiley-Interscience, New York, 1980, Vol. 1, p. 657).

The correct dosage is very important and enough chlorine is needed to achieve the required brightness, too much can result in degradation of the carbohydrates and a reduction in physical strength of the fibres. Chlorination with molecular chlorine tends to substitute the aromatic ring in lignin directly and to give products which are not soluble in water but which are soluble in alkali. Chlorinated lignin also contains a high proportion of acidic groups which makes it amenable to alkaline extraction. Chlorine dioxide is often used because it is less damaging to the carbohydrate fraction than chlorine. It is, however, a rather unstable and very reactive compound, having an unpaired electron, and at concentrations above around 12–15% it is explosive. It tends to react with the aromatic rings in lignin not by substitution but by destruction of its aromaticity to produce charge transfer complexes *via* free radical mechanisms. For this reason, chlorine dioxide produces only about 10% of the amount of organochlorine compounds which are found in chlorine-bleached pulps. Most of the colour removal seems to be due to the opening of the aromatic ring.

## HIGH PURITY DISSOLVING PULPS

An important use of wood pulp, which does not involve paper, is in the preparation of soluble cellulose derivatives such as ethers and esters, and for the preparation of regenerated cellulose fibres such as viscose rayon. These products require a pulp with a high cellulose content with as high a molecular weight as possible and which is free of lignin and low molecular weight hemicelluloses. These dissolving pulps, as they are known, are made by the pulp industry and have a high degree of brightness. They are most commonly prepared by removing the remaining lignin, hemicellulose and resins from hardwoods which have been pulped by the sulfite process. The acidic conditions of the sulfite process are particularly suitable as they remove much of the hemicellulose, which is very sensitive to acid hydrolysis. Alkali-based processes can be used, but a pre-hydrolysis step is normally required. Extensive purification *via* bleaching and extraction with dilute alkali at elevated temperatures and with concentrated alkali at room temperature is then required. The main application of these pulps is in producing soluble derivatives such as ethers (*e.g.* carboxymethyl cellulose) or esters (*e.g.* cellulose acetate) and in the production of cellulose II, a different polymorphic form of cellulose which is obtained by dissolving native cellulose (cellulose I) and regenerating it as a filament or as a film.

# Chapter 4

# Cellulose Fibre Networks

## INTRODUCTION

Paper is a layered fibrous network structure and its mechanical, optical and other properties are therefore highly dependent upon the nature of this network. It is layered in the sense that the fibres lie predominantly in the plane of the sheet and are broadly parallel to each other in the $z$-direction (*i.e.* through the thickness of the sheet). However, the distribution of the fibres in the $x-y$ plane is responsible for the areal mass distribution. Figure 4.1(a) shows a simulated sheet structure arising from the random distribution of 970 straight fibres of uniform length. It is instructive to see the similarity between this and a photomicrograph of a $2.5\,\mathrm{g\,m^{-2}}$ sheet of paper [Figure 4.1(b)] in which the mean fibre length and density correspond to that of the comparable randomly distributed network. Thin sheets made from dilute suspensions can quite closely resemble ideal random structures but commercial processes must use less water and result in more 'clumpy' structures.

The randomly distributed sheet clearly exhibits areas of low and high density but it still remains a good target and a unique reference structure in making paper. The small scale non-uniformities of paper structure are particularly important in their influence on pore size distribution and the distribution of areal mass density, and both of these properties have an influence on mechanical and other properties of the final sheet.

At the point of contact between the cellulosic fibres a strong bond is formed once the fibres have been dried. These bonds are formed by hydrogen bonds between the polysaccharides at the fibre surface. Mechanical and other properties of paper are not only dependent on the nature of the fibre distribution but also on the bonding between the fibres and the inherent fibre strength itself. Bonding between

(a)

(b)

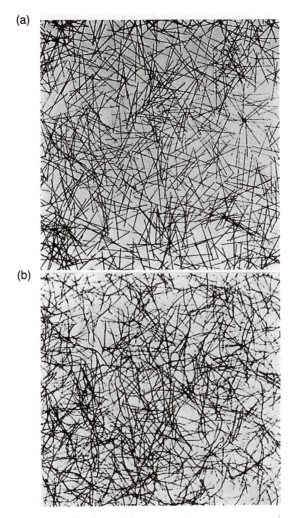

**Figure 4.1** *Sheet structures arising from* (a) *the random distribution of 970 straight fibres of uniform length,* (b) *a photomicrograph of a 2.5 g m$^{-2}$ sheet of paper in which the mean fibre length and density correspond to that of* (a).
(Source: Reproduced from O. Kallmes and H. Corte, *Tappi*, 1960, **43**, 738).

fibres and also fibre strength are influenced by the pulping and bleaching and fibre preparation conditions employed prior to sheet formation. This chapter attempts to describe the structure of paper at both the molecular level of bonding and at a more macroscopic

structural level. The structure of the primary component polysaccharide, cellulose, is discussed in some detail and this is followed by
a discussion of the general structural features of cellulose and their
relationship to physical properties of the sheet.

## THE STRUCTURE OF CELLULOSE

### Molecular Structure

The primary structural component of paper is cellulose but non-
structural polysaccharides (hemicelluloses) and sometimes lignin
may also be present in paper. The physical and mechanical properties of a sheet are, however, in large measure due to the cellulosic
fibres.

Naturally occurring cellulose is a polydisperse linear homogeneous
polysaccharide based on $\beta$-1,4-D-glucopyranose repeat units, with an
average degree of polymerisation variously estimated to be in the
range 3000 to 15 000 depending upon source. Wood cellulose does
not have a particularly high molecular weight and the highest
molecular weight celluloses are generally obtained from non-woody
sources such as flax and cotton. Each D-glucopyranose unit is in the
$^4C_1$ conformation and the chain is in a highly extended conformation
exhibiting two-fold symmetry (*i.e.* every glucose unit is rotated by
approximately 180° relative to its neighbour). Conformational analysis predicts reasonably accurately the two torsion angles around the
glycosidic linkage ($\phi$ and $\psi$) and these are consistent with those
expected for a highly extended chain and similar to those found in
other structural polysaccharides such as cellobiose (Figure 4.2).

The torsion angles predicted by conformational analysis agree
closely with those of crystalline cellobiose as measured by X-ray
diffraction, the conformation of which is restricted by two chain-
stabilising intramolecular hydrogen bonds between $O(3')$–H and
$O(5)$ and also between $O(2')$–H and $O(6)$ (Figure 4.3). These are
also found in cellulose and they assist in maintaining the highly
extended conformation which allows it to function as a structural
polymer.

### Crystal Structure

Cellulose is partly crystalline and partly amorphous, the percentage
crystallinity varying between 50 and 90% depending upon source
and also upon the method of crystallinity measurement. Numerous
theoretical models have been proposed for the molecular organisa-

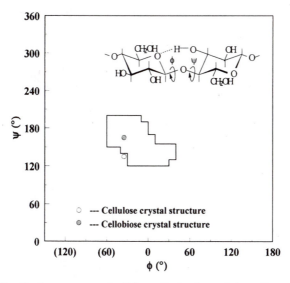

**Figure 4.2** *Conformation map of cellobiose. Enclosed area defines allowed conformations in which there are no major conformational restrictions arising from interactions between non-bonded atoms.*

tion of these amorphous and crystalline regions. In some of these approaches, single cellulose molecules are considered to pass through regions of high and low lattice order, and in others, highly crystalline units are considered to be embedded in amorphous regions. The crystal structure of native cellulose (cellulose I) itself is still an unresolved subject. It is generally accepted to have a parallel chain orientation and to exist in at least two allomorphic forms: I$\alpha$ (monoclinic) and I$\beta$ (triclinic), the proportions of which vary from source to source. The algal and bacterial celluloses are rich in the I$\alpha$ allomorph, and cotton and higher plants are rich in the I$\beta$ form. The $^{13}$C NMR solid-state spectrum can be used to observe the two forms, but the hemicelluloses in wood pulp tend to interfere with spectral definition. The I$\alpha$ form exhibits a C-1 singlet and the I$\beta$ form a doublet. The C-1 signal of mixtures is therefore a linear combinations of these two signals (Figure 4.4).

In addition, cellulose undergoes changes in crystalline structure with relative ease. The most common modification is the conversion of cellulose I (*i.e.* I$\alpha$ and I$\beta$) to cellulose II. This can be achieved by dissolution and regeneration or by simply treating cellulose I with sodium hydroxide. Cellulose II is usually considered to be more thermodynamically stable than biosynthesised cellulose I. However,

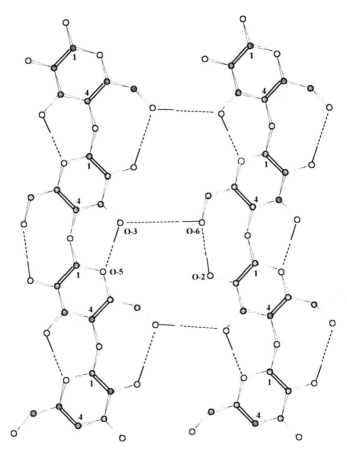

**Figure 4.3** *Inter- and intra-molecular hydrogen bonds of native cellulose.*
(Source: Adapted from 'Handbook of Physical and Mechanical Testing of Paper and Paperboard', ed. R.E. Mark, Marcel Dekker, New York, 1983, p. 413).

the recent *in vitro* synthesis of cellulose I from β-cellobiosyl fluoride must raise doubts about this assumption.

## BONDING IN PAPER

It is clear from our knowledge of material science that the physical properties of materials are dependent upon the nature of the chemical bonding and also upon the type of defects which are present. Paper is a heterogeneous material and its properties are

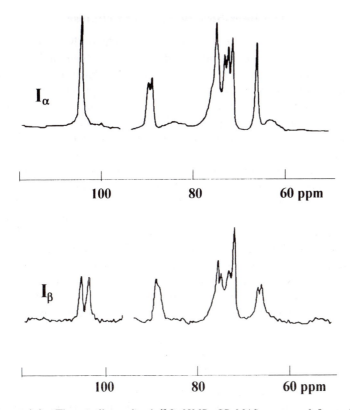

**Figure 4.4** *Theoretically predicted $^{13}C$ NMR CP-MAS spectra of Iα and Iβ cellulose.*
(Source: Adapted from D.L. Vanderhart and R.H. Atalla, *Macromolecules*, 1983, **17**(8), 1465).

dependent both upon the character of the fibres themselves and also upon the characteristics of the bonds between them. In addition to this, there is the added complication of the distribution of the fibres and the bonds within the sheet structure. The continuous filtration nature of the paper-making process, which has been discussed briefly in Chapter 1, results in a layered structure and, because of the flow characteristics of the paper machine, there is also a preferential orientation of the fibres in the direction of flow on to the machine (machine direction) which gives rise to physical anisotropy in the sheet. To complicate matters, during drying the sheet undergoes local shrinkages of as much as 20%.

## The Inter-fibre Bonding of Cellulose

The remarkable property of cellulose fibres which gives rise to their widespread use in paper and board products is their ability, when dried in contact with each other from water, to form a strong bond. Perhaps more importantly, this bond can be completely disrupted by the re-addition of water and this is the essential property which allows cellulosic fibres to be relatively easily recycled.

The bonds between fibres are generally accepted to be due to multiple hydrogen bonds within the bonded area between contacting fibres. Because the bond lengths of hydrogen bonds are of the order of only a few nanometres, the two surfaces must come into very close contact for bonding to occur. Surface tension forces are responsible for bringing the wet fibres together so that this bonding can take place, and these forces become quite large as water is removed from the wet web. Hydrogen bonding is thought to occur as the water removal reaches a point of about 10–25% solids. At about 25% solids, the surface tension forces are dependent inversely upon the thickness of the water film. The pressure difference $\Delta p$ between two surfaces separated by a water film of thickness $x$ is given by:

$$\Delta p = \frac{2\sigma}{x}$$

where $\sigma$ is the surface tension of the water.

Decreasing the water film thickness leads to a very high differential pressure allowing the surfaces to approach close enough for hydrogen bonding to occur. The extent of hydrogen bonding over the area of contact is clearly important, and depends on the ability of the two surfaces to conform to each other. Thus, the flexibility of the fibres in the wet state is an important characteristic and is influenced by the extent of swelling of the fibre cell wall. This point is discussed more fully in Chapter 5.

The nature of the bonds between cellulosic fibres in paper has been the subject of some controversy over many years. The early and now largely discredited view was that paper derived its strength merely from mechanical entanglement of the fibres. However, experiments in which paper is formed from non-aqueous solvents produce sheets with very poor strength properties and have thus tended to disprove this conjecture. In the mid-1950s deuteration experiments were carried out which demonstrated that of the order of 0.4–2% of all hydroxy groups are additionally bonded in paper as compared with the unbonded fibres. This observation led to the view that hydrogen bonding was the primary mechanism for bond formation between cellulosic fibres. However, the precise molecular species involved in hydrogen bonding is a more difficult question. In general, pure cellulosic surfaces such as those found in cotton or bacterial cellulose exhibit rather poor bonding characteristics, whereas fibres derived from wood sources show much better bonding characteristics. This gave rise to the view that adsorbed polysaccharides of the hemicellulose type might also be involved in the formation of inter-fibre hydrogen bonds, or that some form of molecular disruption of the crystalline surface occurred during mechanical action.

## Bonding and Mechanical Strength

Although there is now wide agreement that hydrogen bonding is the primary mechanism of inter-fibre bonding, there is still much dispute over the precise contribution that it makes towards the overall mechanical strength of paper. Two theoretical approaches have been used to explain the mechanical properties of paper. The first considers paper as a continuously hydrogen bonded solid, and the second considers the mechanical strength of paper as being due partly to the inter-fibre bonds and partly to the inherent strength of individual fibres. The latter view has largely prevailed, probably as a result of experiments such as the one shown in Figure 4.5 which demonstrates that the tensile strength of paper is a linear function of the number of fibres which fail during the test.

The ultimate strength of paper may therefore be regarded as that in which 100% of fibres break during failure. This value can be determined experimentally by measuring the tensile strength at zero span. In this test a sample of paper is held between jaws at notionally zero span causing a high proportion of fibres across the line of failure to be fractured during the measurement.

## MECHANICAL STRENGTH

### Directional Anisotropy

Because paper is made from a flowing suspension, fibres tend to be laid down preferentially with their long axis in the plane of the sheet (layered structure), and with the axis aligned broadly parallel to the flow of the paper through the machine. In addition, there is also some web tension and drying restraint which gives rise to an

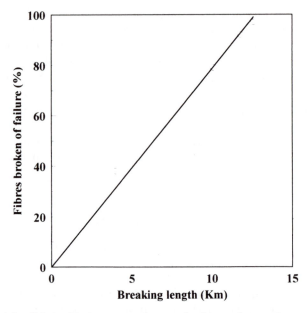

**Figure 4.5**   *Relationship between tensile strength of paper (expressed as a breaking length) and the number of fibres which fail during the test.*
(Source: T. Helle, *Svensk Papperstidning*, 1963, **66**(24), 1015).

orthotropic material response. Three mutually perpendicular directions may therefore be identified: the machine direction (MD), the cross machine direction (CD) and the thickness (or $z$ direction). These are shown schematically in Figure 4.6.

Paper made on a paper machine exhibits quite different properties in the $x$ and $y$ directions (the machine and cross machine directions), an example of which is a difference in stiffness which can be demonstrated by plotting the specific elastic stiffness in the $x$–$y$ plane as a function of the machine direction and cross machine direction co-ordinates in the form of a polar diagram (Figure 4.7).

The area of the polar diagram is related to variables such as refining and wet press pressure. The load–elongation curve during tensile testing also shows marked differences in the two directions (Figure 4.8).

It is obvious that the response to tensile forces applied on each of the principle directions will be different due to the orientation of the fibre segments and of the bonded areas connecting them. Forces acting on a hypothetical diamond-shaped portion of the fibre net-

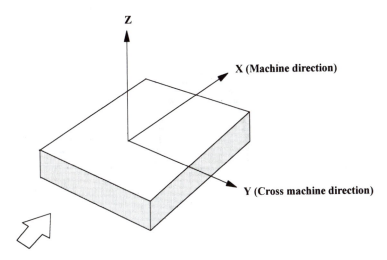

**Direction of paper flow**
**through paper machine**

**Figure 4.6** *The machine direction (MD), cross machine direction (CD) and thickness direction (z) of a sheet of paper.*
(Source: Adapted from 'Handbook of Physical and Mechanical Testing of Paper and Paperboard', ed. R.E. Mark, Marcel Dekker, New York, 1983, p. 157).

work are clearly different when loaded in the machine and cross machine directions (Figure 4.9).

## The Effect of Moisture on Mechanical Strength

One of the most important problem areas in many commercial paper products is the loss of strength which occurs on increasing the moisture content. In products such as tissue, sacking, wall paper, etc., it is necessary to use chemical means to enhance artificially the so-called 'wet strength' of paper, and this subject is discussed more fully in Chapter 7. Water competes for sites of hydrogen bonding between fibre surfaces and thus reduces the strength of inter-fibre bonding leading to a sharp change in the load–elongation curve at different moisture contents (Figure 4.10).

## Paper Formation

It is important that high quality grades of paper, such as those used for printing, writing and artwork, should have as uniform a fibre

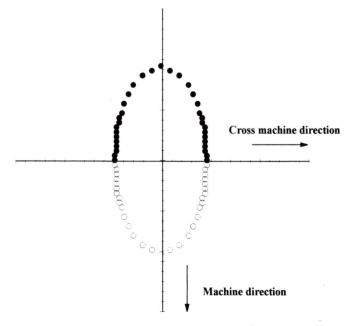

**Figure 4.7**   *Difference in rigidity between the machine and cross machine directions.*
(Source: 'Subfracture Mechanical Properties', G.A. Baum, in
'Products of Paper Making', Trans. 10th Fundamental Re-
search Symp., PIRA International, 1993, p. 53).

distribution as possible. Paper makers refer to this property as the
formation of the sheet, and it manifests itself visually as a variable
distribution of opacity of the sheet in the $x–y$ plane which makes the
sheet appear 'blotchy' when held to the light. It is caused by flocs of
fibres forming in the fibre suspension during the sheet-forming
process and the problem is more severe when long fibres are used as
these have a greater tendency to mechanical entanglement. The
problem is also exacerbated by the use of polymeric flocculants as
retention aids (this subject is discussed more fully in Chapter 7).
More correctly, formation is a variance in mass distribution in the
$x–y$ plane and this is the basis of quantitative methods for its
measurement. Examples of poorly and well-formed sheets of paper
and also the fibre suspension from which these were made are shown
in Figure 4.11.

Formation is important not only because of its impact on the
aesthetic appearance of the sheet but also because it adversely affects
both mechanical and optical properties. A more uniform mass

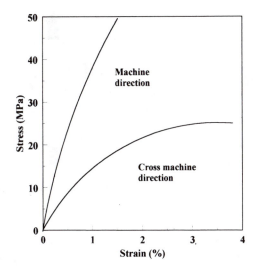

**Figure 4.8** *The load–elongation curves for paper in the machine and cross machine directions.*
(Source: 'Handbook of Physical and Mechanical Testing of Paper and Paperboard', ed. R.E. Mark, Marcel Dekker, New York, 1983, p. 181).

distribution would lead to a sheet of higher tensile strength and higher opacity. The relationship between opacity and grammage (mass per unit sheet area) is not a linear one and therefore care must be taken in interpretation of the relationship between formation and mechanical properties, when data is used which is based on transmission measurements.

## Descriptive Models of Paper Strength

Two types of models have been applied to the mechanical strength of paper. The first assumes paper to be a continuous network of hydrogen bonds with no other type of bond contributing to its mechanical properties, and the second describes its mechanical strength in terms of a combination of fibre strength and fibre-to-fibre bonds.

In the first approach, Young's modulus is related to the number and strength of effective hydrogen bonds taking part in storing the mechanical energy during any axial straining per unit volume of sample, and the model behaves reasonably well in describing the weakening effect of paper which arises from an increase in both

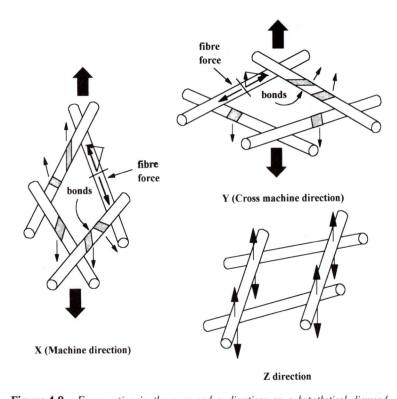

**Figure 4.9**   *Forces acting in the* x, y *and* z *directions on a hypothetical diamond-shaped portion of a fibre network.*
(Source: Adapted from 'Handbook of Physical and Mechanical Testing of Paper and Paperboard', ed. R.E. Mark, Marcel Dekker, New York, 1983, p. 157).

temperature and moisture content. However, only the very early stages of straining are similar in all papers and it eappears that, in the later stages of straining, the hydrogen bond model works less well, and structural considerations become more important. The structural approaches assume that a combination of fibre strength and fibre-to-fibre bonds are responsible for the mechanical strength of paper but, whilst fibre strength is relatively easy to determine, the bonding strength is more difficult. It is generally accepted that, for fibre-to-fibre bonding to occur, the fibres must be in close optical contact. The relative bonded area (RBA) may be defined as the proportion of the total surface area in optical contact. For two fibres of length $\lambda$ and width $\omega$ the total surface area if it is assumed that they are flat rectangular ribbons is $4\lambda\omega$. If they overlap at right

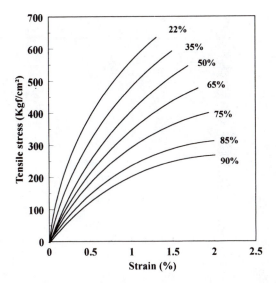

**Figure 4.10** *The effect of relative humidity (%) on the stress–strain curve of paper.* (Source: 'Handbook of Physical and Mechanical Testing of Paper and Paperboard', ed. R.E. Mark, Marcel Dekker, New York, 1983, p. 169).

angles, the total area in contact is $2\omega^2$. If a fraction, $\beta$, of this contacting area is in close optical contact, then the relative bonded area is given by:

$$\text{RBA} = \frac{2\omega^2\beta}{4\lambda\omega}$$

The experimental determination of RBA, however, is difficult but some attempts have been made and these include direct observation, measurements of electrical conductivity, shrinkage energy, gas adsorption and light scattering. The linear elastic response of paper has been explained in terms of various micromechanical models which take into account both fibre and network properties, including RBA. An example of one which predicts the sheet modulus, $E_s$ is given below:

$$E_s = (\tfrac{1}{3})E_f[1 - (w/L.\text{RBA})(E_f/2G_f)^{1/2}\tanh\{(L.\text{RBA}/w).(2G_f/E_f)^{1/2}\}]$$

where $E_f$ and $G_f$ are the fibre elastic and shear modulii, $w$ and $L$ are the mean fibre width and 'effective' length.

**Figure 4.11**   *Flocculation and formation effects in a chemically pulped bleached softwood pulp (slightly refined). Fibre suspensions settled for 40 min. Sheets (60 g m⁻²) photographed in transmitted light: (a) no additives, (b) polyelectrolyte added to induce flocculation. Scale bar = 2 cm.*

## LIQUID PENETRATION INTO PAPER

The penetration of fluids into paper is also a very important material property for many product types. It is influenced by a number of factors, not least of which is the sheet structure and porosity. In some cases these can be more dominant than the surface energy of the component fibres. The sheet structure can be controlled to a large extent by the selection and refining of the pulp. Structural considerations notwithstanding, the absorption of fluids by paper should be considered to be a combination of both surface wetting and capillary pore penetration.

If, when a liquid drop is placed on a smooth surface, the forces of adhesion between the solid and the liquid are greater than the forces of cohesion of the liquid, then the liquid will spread and will perfectly wet the surface spontaneously. If the forces reach an intermediate balance determined by the interfacial energies $\gamma_{lv}$, $\gamma_{sl}$ and $\gamma_{sv}$, then the liquid drop will form a definite contact angle $(\theta)$ with the solid surface (Figure 4.12).

This wetting process may be described in terms of a balance of specific surface energies — the Young equation:

$$\cos \theta = \frac{\gamma_{sv} - \gamma_{sl}}{\gamma_{lv}}$$

where $\gamma_{sv}$, $\gamma_{sl}$ and $\gamma_{lv}$ are respectively, the solid–vapour, solid–liquid and liquid-vapour interfacial energies. However, paper is a porous material and when a liquid contacts a porous solid, the liquid in contact with the pore becomes curved due to differential surface tensions. For a pore of cylindrical cross-section, the pressure difference, $\Delta p$, across the curved surface may be expressed in terms of the contact angle, $\theta$, the liquid–vapour interfacial tension, $\gamma_{lv}$, and the radius of the cylindrical pore, $r_c$:

$$\Delta p = \frac{2\gamma_{lv} \cos \theta}{r_c}$$

Clearly, if the contact angle between the solid and the liquid is

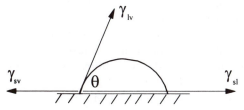

**Figure 4.12**  *Contact angle (θ) formed by a liquid droplet in contact with a solid surface.*

greater than 90°, $\Delta p$ is zero and the liquid will not penetrate by capillary action. However, this equation defines an equilibrium position, and paper makers are more concerned with the dynamic process of penetration. The dynamic rate of capillary suction of fluids into paper has been effectively described by models of penetration of fluids into a single capillary (the Washburn equation). This gives a good approximation to the rate of capillary intrusion of non-swelling fluids into paper. Modifications of the basic equation are necessary, however, to describe the behaviour of swelling fluids such as water:

$$\frac{\mathrm{d}l}{\mathrm{d}t} = \frac{\gamma_{\mathrm{lv}} r \cos \theta}{4\eta l}$$

where $l$ is the distance penetrated into a cylindrical capillary of radius $r$ in time $t$ by a liquid of surface tension $\gamma_{\mathrm{lv}}$ and viscosity $\eta$.

Retardation of the rate of penetration is necessary for many products and this can be brought about by the creation of a low energy, hydrophobic surface at the fibre–water interface which increases the contact angle formed between the drop of liquid and the surface. This important change can be achieved chemically in the process known as sizing which is discussed more fully in Chapter 7.

If the surface of paper has been modified by coating or by the application of a film-forming polymer, there will be a relatively dense layer of material on the surface of the sheet through which the test fluid will have to penetrate. This will cause a significant difference in the rate at which a test fluid will penetrate the surface layers of the web and the rate at which it penetrates the inner layers of the web. The size of the pores in the paper are also important. An aqueous test fluid can penetrate a sheet of paper either *via* the pores in the sheet (the areas between the fibres in the web) or *via* the fibres. The larger the average pore size in a given sheet of paper, the greater will be the probability that the fluid will penetrate the sheet *via* the pores rather than the fibres.

If the fibres in the paper have been rendered hydrophobic by sizing, but the sheet has an open structure and there has been no surface treatment to cover the sized fibres, then the web will show a high contact angle. However, if the same web is tested by a penetration-type test, the sizing level will be low.

*Chapter 5*

# The Paper Formation Process

## INTRODUCTION

Once the wood has been converted into pulp by either mechanical or chemical means, it is then ready to be formed into paper. This may be done on the same site as the pulping operation, in what is known as an 'integrated mill', or the pulp may be dried and transported to another manufacturing site for subsequent processing. The conversion to paper involves many steps, each of which has an impact upon the properties and character of the final product. For example, it is possible by manipulation of either or both of the chemical and mechanical processes to produce products as wide ranging as grease-proof paper to highly absorbent tissue. There are three important stages in the treatment of the pulp prior to its delivery to the paper machine, and these are known collectively as stock preparation. The first is the dispersion of the pulp as a slurry in water (this is not necessary for an integrated process in which the pulp has never been dried), the second is the mechanical refining or beating of the fibres to develop appropriate physical and mechanical properties for the product being made, and the third is the addition of chemical additives which impart specific product properties or facilitate the paper-making processes itself. After these processes, the pulp suspension is then usually diluted and flows on to the paper machine for the paper forming part of the process. The chemical aspects of these stages are discussed in this chapter.

## FIBRE PRETREATMENT

### Dispersion

This process is somewhat ambiguously referred to as pulping, but should not be confused with the chemical delignification process which is also referred to as pulping. The dispersion of the fibre in water may be carried out as a continuous or as a batch process, and it is common for different pulp types (*e.g.* chemical or mechanical pulp, recycled fibre) to be processed separately and then mixed at a later stage. The purpose of this process is to ensure the total and separate dispersion of individual fibres of the pulp sheets into an aqueous suspension. The equipment employed usually consists of a large circular tank with a revolving rotor at the base to provide the turbulence and circulation necessary to disintegrate the fibres. Rotor design has changed substantially in recent years to facilitate dispersion at higher consistency (as high as 15 to 18%). The pulp bales are normally fed into the top of the open pulper tank. It is not the purpose of this book, however, to discuss the detailed mechanics of this process and the reader is directed to other texts in this area (*see* Recommended reading).

## Refining

This process, when conducted as a batch operation, is known as 'beating', and the two terms 'refining' and 'beating' are sometimes used synonymously. It is common these days to consider refining as a continuous operation and beating as a batch operation, however, the two processes in terms of their mechanical effect upon the fibres are essentially the same. Details of the mechanical design of beaters and refiners can be found elsewhere, and the purpose of this chapter is to discuss the physical and chemical effects of this process on the fibre and also its effect upon ultimate sheet properties.

The refining process involves the circulation of the fibre suspension in such a way as to force the fibres between a stationary metal plate (the stator) and a moving metal plate (the rotor). As the fibres are wet at this stage, both mechanical and hydraulic forces are involved in altering fibre characteristics. Both shear and normal stresses (either tensional or compressive) are imposed on the fibres in this process, and the mechanical action is shown diagramatically in Figure 5.1.

Refining is the most important of all the processes to which fibres are subjected, in terms of developing pulp suspension characteristics and final sheet properties, and a great deal of research has been carried out into understanding the process more fundamentally. Whilst there is still much controversy about certain aspects of the refining process and its effects upon the fibres, a number of things are widely accepted. Firstly the primary cell wall, which does not

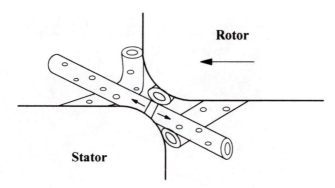

**Figure 5.1** *Schematic representation of the mechanical action of the refiner.*
(Source: Adapted from 'Handbook for Pulp and Paper Techno-
logists', G. Smook, Angus Wilde Publications, Vancouver,
1992, p. 200).

normally swell easily and therefore tends to prevent the rest of the
fibre from swelling, is partially removed, thus allowing the secondary
cell wall to become exposed and at the same time allowing water to
be absorbed into it. The swelling which takes place at this stage
causes the fibres to become soft and more flexible ('internal fibrillation'). In addition, some of the microfibrillar structural components
of the cell wall are loosened from the surface giving rise to a very
large increase in the surface area of the fibre ('external fibrillation').
As the fibre becomes more flexible, the cell walls, on drying, tend to
collapse into the lumen, giving a more ribbon-like structure. The
effects of this process can be seen more clearly in Figure 5.2 which
show the changes in individual fibres as a result of refining.

There is inevitably some fibre shortening during the refining
process which is caused by the shearing action on the fibres during
their passage between the rotor and stator. This tends to reduce
strength and contributes to slower drainage, and this has implications for the maximum speed at which the sheet can be made.
However, fibre shortening is not always undesirable, and the mass
distribution of the fibre within the sheet is generally improved by it.

After refining, the fibre suspension is then cleaned to remove
particles of grit, *etc*. This is essentially a mechanical process which
involves no chemistry and is not discussed further here. Refining on
the other hand has a very influential effect on the fibres. The
changes in fibre swelling, length and flexibility give rise to important
changes in both pulp and paper properties and these are now
discussed in more detail.

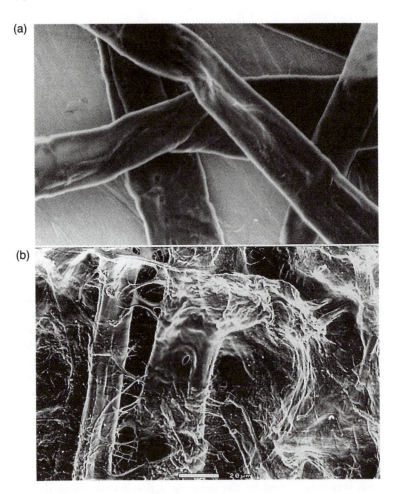

**Figure 5.2**   *Environmental scanning electron photomicrographs of fibres of a chemically pulped (sulfite process) softwood* (a) *before refining and* (b) *after
refining.*
*Scale bar = 20 μm.*

## THE EFFECTS OF REFINING

### Pulp Surface Area

The surface area of fibres increases during refining. However, the
definition of surface area for cellulose is not straightforward and it is
necessary to define it a little more precisely. It can be measured
when the pulp is in either the wet or the dry state and very different

values are obtained. The method of sample preparation thus affects results dramatically. Attempts to preserve the water-swollen state (*e.g.* by solvent exchange) give much higher results than for fibres dried from water where no attempt has been made to preserve the water-swollen state. Surface area measurements of the latter give values between 0.5 and 3 $m^2 g^{-1}$, whereas those of the former lie in the range 100 to 150 $m^2 g^{-1}$. It is has been established by X-ray diffraction that water does not penetrate the crystalline regions of cellulose (the X-ray diffraction spacing of the crystalline lattice is independent of relative humidity) and refining is therefore considered to cause water to enter the cell wall only in the amorphous regions between crystalline zones ('intercrystalline swelling'). Some solvents are known to swell the crystalline zones ('intracrystalline swelling') but, although this is important in other areas of cellulose science and technology, it is not particularly relevant to paper making which is done in an exclusively aqueous environment. The actual method of determining the surface area also causes some wide variation. The methods of measurement may be chemical (for example by reaction of accessible OH groups) or physical [usually $N_2$ adsorption using the Brunauer–Emmett–Teller (BET) adsorption isotherm]. Some comparison of surface area measurements for cellulose samples prepared in the same way by $N_2$ adsorption and by thallation (*i.e.* by reaction of available OH groups with thallium (I) ethylate in benzene then replacement of thallium by methyl groups using MeI followed by measurement of MeO content) are shown in Table 5.1.

It is also possible to measure surface area from the water adsorption isotherm, and this is arguably more relevant to aqueous pulp suspensions as it measures the surface area which is accessible to water. Values of up to 140 $m^2 g^{-1}$ have been obtained from the

**Table 5.1** *Effect of swelling on the surface area of cotton as measured by chemical (thallation) and physical ($N_2$ adsorption) methods.*
(Source: G.A. Roberts, 'Accessibility of Cellulose', in 'Paper Chemistry', ed. J.C. Roberts, ch. 2, Blackie, Glasgow, 1991).

| Fibre type | Surface area ($m^2 g^{-1}$) | |
| --- | --- | --- |
| | Thallation | $N_2$ Adsorption |
| Cotton fibre | 16 | 0.55 |
| Cotton fibre swollen in water | 263 | 137 |

analysis of water vapour adsorption isotherms which reflect the water-swollen state of the fibres. It is difficult, however, to distinguish between 'adsorption' and 'absorption' processes when pulps interact with water. Water is first adsorbed monomolecularly by hydrogen bonding to accessible OH groups of the cellulose (*i.e.* those not engaged in holding the crystalline matrix together) and, with increasing adsorption, the degree of coverage of the monolayer increases. Finally, several layers are adsorbed successively and, as the thickness of the layers increases, the size of the pores increases (amorphous regions). At high vapour pressures, water is gradually condensed in the capillaries.

Several theoretical models have been proposed to explain the water adsorption isotherm. These are based either upon adsorption (of the BET type) or upon solution theories (for example, using the Flory Huggins equation). The most widely used of these has been the BET approach, which extends the monomolecular Langmuir-type adsorption of gases on surfaces to include multilayer adsorption. Given a knowledge of the effective cross-section of the adsorbing molecule it is possible to calculate the surface area of the adsorbent. Although the BET theory has been criticised for ignoring adsorbate interactions and also the heterogeneity of the substrate, surface area values have been obtained which are in good agreement with other methods. Some examples are given in Table 5.2.

Water which is bound to cellulose (or any other natural polymer) has properties different from those of unbound (bulk) water. For example, it has a higher density and a lower freezing point. The

**Table 5.2** *Comparison of specific surface areas calculated from the ascending branch of the water isotherms of selected cellulosic materials by the BET theory, the 't' procedure and the Zimm and Lundberg (ZL) cluster theory.*
(Source: T.P. Nevell and S.H. Zeronian, 'Cellulose Chemistry and its Applications', p. 147, Ellis Horwood, Chichester, 1985).

| Sample | Surface area ($m^2 g^{-1}$) | | |
|---|---|---|---|
| | *BET* | *t* | *ZL* |
| Cellophane | 216 | 233 | 216 |
| Filter paper | 128 | 128 | 113 |
| Delignified maple holocellulose | 172 | 177 | 179 |
| Delignified aspen holocellulose | 234 | 202 | 201 |

density increase can be explained by the fact that accessible OH groups strongly attract the dipoles of the water molecules, and the water molecules will therefore be oriented close to these hydroxy groups and will consequently lose some of their mobility. They will therefore be packed more efficiently and their density will exceed that of free water. Subsequent molecules will not be so strongly bound, and will have a higher mobility and, hence, a lower density. The density of water thus decreases with increasing distance from the surface.

## Water Sorption

Sorption of water vapour by cellulose is an exothermic process but, because it is a slow process and the energy is easily dissipated, it is difficult to measure quantitatively. However, the heat of wetting, $\Delta H_w$ ($J g^{-1}$), can be measured calorimetrically by submerging cellulose which has been equilibrated in air in an excess of water and measuring the temperature rise. The total heat of wetting ($\Delta H_w^\circ$) is the amount of energy released when a completely dry sample is submerged in water. Some examples of cellulosic heats of wetting properties are given in Table 5.3.

The relationship between the heat of adsorption and the heats of wetting is shown in Figure 5.3. $\Delta H_w^\circ$, the total heat of wetting, must equal the pure heat of adsorption of water ($\Delta H$) and the energy which is consumed to wet the sample ($\Delta H_w$):

$$\Delta H_w^\circ = \Delta H + \Delta H_w$$

Hence, $\Delta H$ can be calculated from $\Delta H_w^\circ$ and $\Delta H_w$.

**Table 5.3** *Total heats of wetting* ($\Delta H_w^\circ$) *of some cellulose samples.* (Source: D. Eklund and T. Lindstrom, 'Paper Chemistry—An Introduction', DT Paper Science Publications, Grankulla, Finland, 1991, p. 36).

| Type of cellulose | Total heat of wetting ($J g^{-1}$) |
|---|---|
| Chemical pulps | 53–58 |
| Cotton | 44.1 |
| Viscose | 106 |
| Ramie | 44 |
| Spruce wood meal | 83 |

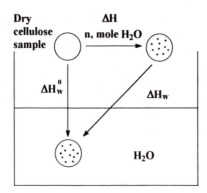

**Figure 5.3**  *Relationship between heat of adsorption and heats of wetting.*
(Source: Adapted from 'Paper Chemistry—An Introduction',
eds. D. Eklund and T. Lindstrom, DT Paper Science Publications, Grankulla, Finland, 1991, p. 37).

Adsorption of water by cellulose displays hysteresis. The adsorption isotherm is not identical to the desorption isotherm and the
amount of adsorbed water in equilibrium with the atmosphere at a
particular relative humidity is higher during desorption from a
higher humidity than during adsorption from a lower humidity. A
plot of the adsorption/desorption isotherm is shown in Figure 5.4.

Hysteresis is observed not only in the sorption isotherms but also
in calorimetric measurements of heat of wetting at different moisture
contents, and it is thus a combined entropy and enthalpy phenomenon. A reliable explanation for this effect is not currently available,
but there is speculation that it is due to the stresses which are
induced as the cellulose swells. Since the swelling of cellulose is not
completely reversible, mechanical recovery is incomplete and hysteresis will therefore be present both in the internal stress–strain
curve of the sample, and also in the water adsorption isotherm.

As water swells cellulose in an intercrystalline way (*i.e.* only
within the non-crystalline amorphous regions), a relationship would
be expected between accessibility and moisture uptake, and this is
indeed found (Figure 5.5). Refining causes cellulosic fibres to swell
and it would therefore be expected to cause a change in the water
adsorption isotherm. This is indeed observed (Figure 5.6).

The surface area increase during refining is also easily demonstrated experimentally. The BET adsorption isotherm, for example,
shows that there is an approximately 250% increase in specific

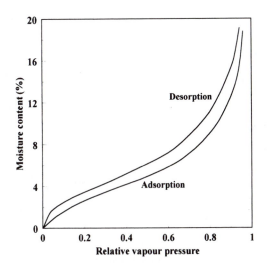

**Figure 5.4** *Water adsorption and desorption isotherms for cellulose.*
(Source: Adapted from 'Paper Chemistry—An Introduction',
eds. D. Eklund and T. Lindstrom, DT Paper Science Publications, Grankulla, Finland, 1991, p. 41).

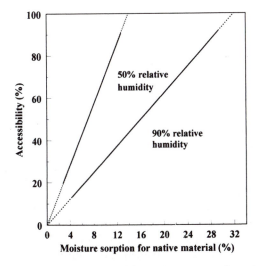

**Figure 5.5** *Relationship between accessibility and moisture uptake at different relative humidities.*
(Source: 'Paper Chemistry—An Introduction', eds. D. Eklund
and T. Lindstrom, DT Paper Science Publications, Grankulla,
Finland, 1991, p. 46).

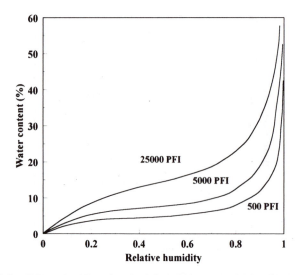

**Figure 5.6** *Effect of refining (number of revolutions in a PFI mill) on the water adsorption isotherm of an unbleached pulp.*
(Source: L.T. Qiang, U. Henriksson, J.C. Eriksson and L. Odberg, in Trans. 9th Fundamental Research Symp., eds. C.F. Baker and V. Punton, Mech. Eng. Publications, London, 1989, p. 45).

surface area going from pulp which has been refined for 500 revolutions in a PFI mill (a small scale laboratory refiner) to pulp which has been refined for 25 000 revolutions. This surface area change can also be observed by deteurium NMR experiments. Figure 5.7 shows the $^2H$ NMR spectrum for $D_2O$ adsorption by sheets which have been orientated perpendicularly to the magnetic field, $B_0$. The spectrum displays a quadrupole doublet centred at the Lamor frequency, and the magnitude of the splitting gives information about the time-averaged orientation of the absorbed water molecule with respect to the magnetic field. The quadrupolar splitting varies as a function of relative humidity for pulps which have been refined to different extents (Figure 5.8).

In addition to the use of quadrupolar splitting, the spin relaxation rate can also be used to calculate the specific surface area ratios for pulps beaten to different degrees and the results for an unbleached pulp agree closely and confirm the 250% increase in surface area measured by isotherm data (Table 5.4).

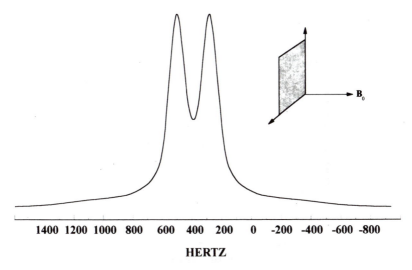

1400 1200 1000 800 600 400 200 0 -200 -400 -600 -800

**HERTZ**

**Figure 5.7** *The ²H NMR spectrum of D₂O adsorbed on pulp sheets oriented perpendicular to the magnetic field.*
(Source: L.T. Qiang, U. Henriksson, J.C. Eriksson and L. Odberg, in Trans. 9th Fundamental Research Symp., eds. C.F. Baker and V. Punton, Mech. Eng. Publications, London, 1989, p. 49).

## Changes in Internal Structure of the Cell Wall

The most important change occurring during refining in terms of its effect upon paper properties is the change in the internal structure of the cell wall. The combined mechanical and hydraulic forces cause the cell wall to delaminate and create voids in which water can be accommodated. The overall process is referred to loosely as the swelling of the cell wall, but the nature and form of the water and also the nature and form of the pores and cracks have been the subject of considerable discussion and controversy. Swelling confers a greater degree of flexibility on the fibres, thus allowing them to conform better in the final sheet. This can easily be observed by comparing sheets made from wet-laid fibres to those formed from fibres which have been laid down in their dry state.

The changes in fibre flexibility may affect not only the rigidity of the fibre but also the local plasticity of the cell wall, and this may be important in determining the ease with which the inter-fibre hydrogen bonds are formed during drying. Pulp types also differ in

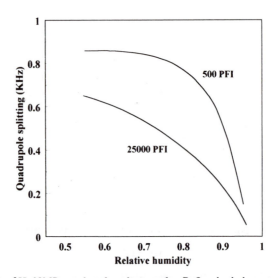

**Figure 5.8**   *²H NMR quadrupole splittings for D₂O adsorbed on to never dried bleached Kraft softwood pulp refined to different degrees (25000 and 500 revolutions in a PFI mill).*
(Source: L.T. Qiang, U. Henriksson, J.C. Eriksson and L. Odberg, in Trans. 9th Fundamental Research Symp., eds. C.F. Baker and V. Punton, Mech. Eng. Publications, London, 1989, p. 51).

**Table 5.4**   *A comparison of specific surface area ratio calculated from quadrupole splitting ($\Delta_Q$), spin-lattice relaxation rate ($R_1$), half-height linewidth ($\Delta v_{1/2}$) and isotherm data for an unbleached linerboard pulp beaten to various degrees.*
(Source: L.T. Qiang, U. Henriksson, J.C. Eriksson and L. Odberg in Trans. 9th Fundamental Research Symp., eds. C.F. Baker and V. Punton; Mech. Eng. Publications, London, 1989, Vol. 1, pp. 39–65).

| Degree of beating (revolution in a PFI mill) | Relative surface area from | | | |
| --- | --- | --- | --- | --- |
| | $\Delta_Q$ data | $R_1$ data | $\Delta v_{1/2}$ data | Isotherm data |
| 500 | 1.0 | 1.0 | 1.0 | 1.0 |
| 25 000 | 2.3 | 2.4 | 2.4 | 2.50 |

their degree of swelling, sulfite pulps swell more rapidly than sulfate pulps, and this can easily be visualised (Figure 5.9).

The initial adsorption of water by fibres up to a relative humidity of 80% is only around 10 to 15%, but paper makers are concerned

(a)

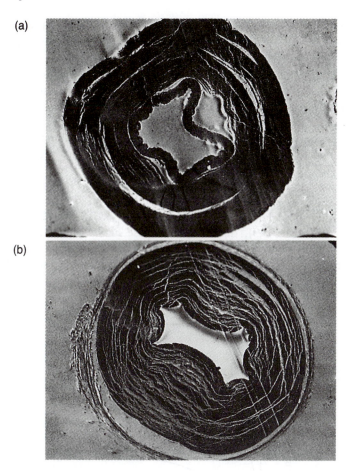

(b)

**Figure 5.9**  *Cross section of* (a) *a Kraft pulp and* (b) *a sulfite pulp, demonstrating the greater delamination of the cell wall of the latter.*
(Source: Reproduced from D.H. Page, in Trans. 9th Fundamental Research Symp., eds. C.F. Baker and V. Punton, Mech. Eng. Publications, London, 1989, p. 18).

with fibres in aqueous suspension where, because of water condensation in the capillary pores of the cell wall, much higher levels of water uptake are observed. It is this water which has always been considered by paper makers to be of vital importance both to the performance of the fibre suspension during the sheet formation process—particularly drainage—and also to final sheet properties. Various techniques have been designed for its measurement.

The two methods most popularly used for measurement of the cell wall water content are the water retention value and the fibre saturation point. In the former the water content of the fibres is determined after removal of interstitial water by centrifugation of a wet sample, whereas in the latter the principle of solute exclusion from the cell wall of a high molecular weight polymer is used. In the latter procedure, a wet sample of pulp is introduced to a known mass of a non-adsorbing high molecular weight polymer solution of known concentration. The cell wall water is not available for dilution of the polymer solution because of the exclusion of the large polymer from the cell wall micropores. The amount of cell wall water can therefore be determined by measuring the change in polymer concentration.

During the delamination of the cell wall which takes place during refining (Figure 5.9), there is an increase in the water accommodated within non-crystalline zones, and the cell wall water content typically rises from around 1 to 3 g per g of dry fibre (Figure 5.10).

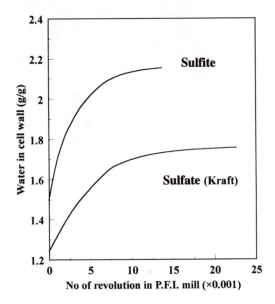

**Figure 5.10**   *The effect of refining on the cell wall water content of a Kraft and a sulfite pulp.*
(Source: Adapted from A.M. Scallan, in 'Fibre−Water Interactions in Paper Making', in Trans. 6th Fundamental Research Symp.', Technical Division of British Paper and Board Federation, London, 1977).

## Sheet Properties

A sheet of paper made from unrefined fibres would have poor tensile strength and poor burst characteristics. It would also be poorly bonded, bulky, absorbent to fluid and have a high porosity and opacity. It would in fact have a very open and irregular structure probably with relatively poor formation (mass distribution). A sheet of paper made from refined fibres, on the other hand, has much greater mechanical strength, high density, low opacity, a smoother surface and a more regular formation. The resistance of the paper to tearing however decreases. These property changes are summarised in Figure 5.11.

A very large number of changes take place within the whole fibre suspension during the refining process and a precise interpretation of the influence of the process upon mechanical and other sheet properties therefore becomes extremely difficult. The more important effects can be summarised as follows. There is a shortening of the fibres arising from a cutting effect and part of the fibre cell wall is

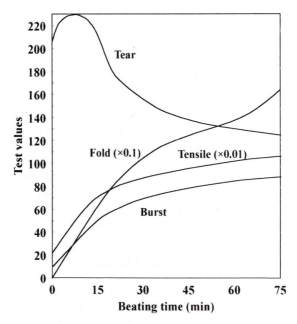

**Figure 5.11** *The effect of beating on various paper properties.*
(Source: Adapted from 'Handbook for Pulp and Paper Technologists', G. Smook, Angus Wilde Publications, Vancouver, 1992, p. 7).

removed, giving rise to debris known as 'fines' in the suspension. Partial removal of the fibre cell wall takes place without complete detachment from the fibre (external fibrillation) and delamination of the internal cell wall structure also occurs allowing an increased uptake of water (internal fibrillation). There is also a change in the degree of curl of the fibres and also changes in the number of nodes, kinks, slip planes and microcompressions in the cell wall. Some dissolution of partially soluble or colloidal material into the bulk solution also takes place and there is a redistribution of soluble polysaccharides from the cell wall to other surfaces. Some of these changes can be considered in a little more detail.

## Fibre Shortening

Before the advent of wood pulps, long fibres such as cotton were used and cutting was an important aspect of the beating operation and was used to reduce the fibre length. Although this still happens to some extent during beating and refining, fibre shortening now tends to be achieved by blending long-fibred softwood with short-fibred hardwood pulps, thus obviating the need for any cutting effect. There is some evidence from the appearance of the cuts in fibres during refining that the fibres fail in a tensile mode rather than by a scissor-like action.

## Generation of Fines

Fragments of the primary and secondary cell wall are generated by the shearing action of the refiner bars, and pulps differ greatly in their tendency to produce these fines. Unbleached sulfate (Kraft) pulps, for example, are much more resistant to the removal of the primary and secondary cell wall than bleached sulfite pulps (Figure 5.12), which may be attributed to the more degrading effect of sulfite pulping and subsequent bleaching on the S1 layer.

## Curling of Fibres

Fibres become curled (Figure 5.13) when they are subjected to high shear and when refining is carried out at relatively high consistency. This curling appears to be due to the repeated flexing and bending of the fibres beyond their yield point. Low yield chemical pulps (*i.e.* those with high levels of delignification) curl more easily than high yield pulps. The former also retain their curled state, whereas high yield pulps tend to straighten spontaneously under fairly mild

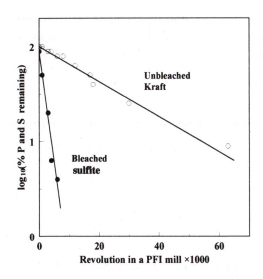

**Figure 5.12** *The removal of the primary (P) and outer secondary (S1) cell wall during refining.*
(Source: D.H. Page, in Trans. 9th Fundamental Research Symp., eds. C.F. Baker and V. Punton, Mech. Eng. Publications, London, 1989, p. 12).

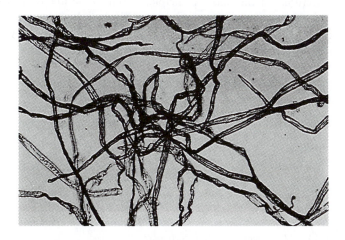

**Figure 5.13** *Curled fibres.*
(Source: Reproduced from D.H. Page, in Trans. 9th Fundamental Research Symp., eds. C.F. Baker and V. Punton, Mech. Eng. Publications, London, 1989, p. 21).

conditions of agitation. When pulp fibres become curled, the sheet becomes more bulky and less dense, the tear resistance of the sheet is improved, and there is a reduction in tensile strength. In other words, the curling of the fibres tends to work in an opposite direction to beating.

## Creation of Dislocations and Microcompressions

Fibres which are refined at high consistency (around 20%) experience repeated bending and axial compressive stresses throughout their length, which give rise to microcompressions within the cell wall. These microcompressions may reduce the apparent fibre length by as much as 5% (Figure 5.14) and they tend to influence the degree of extensibility and dimensional stability of the final sheet.

## THE SHEET-FORMING PROCESS

Once the pulp fibres have been refined to the necessary degree, they are then formed into a sheet of paper on the paper machine. The paper formation process itself is essentially a fast filtration process and involves the delivery of a dilute fibre suspension in water on to a woven endless plastic wire belt, through which it drains to form a wet fibre network. The Fourdrinier paper machine is the most well-established system for forming the wet web, but there are now many variations of this basic principle. A schematic diagram of the Fourdrinier formation process is shown in Figure 5.15.

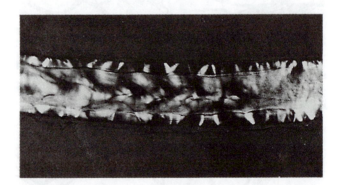

**Figure 5.14**   *Microcompressions in a single fibre.*
(Source: Reproduced from D.H. Page, in Trans. 9th Fundamental Research Symp., eds. C.F. Baker and V. Punton, Mech. Eng. Publications, London, 1989, p. 21).

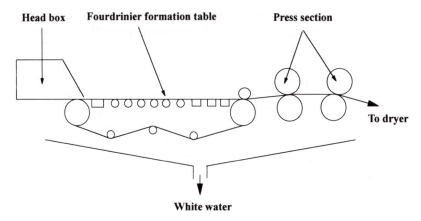

**Figure 5.15** *Paper-forming section of a paper machine.*

The fibre suspension is pumped to the head box of the paper machine which, in modern machines, is usually under pressure; it is discharged from the head box through a narrow orifice extending across the width of the paper machine, known as the slice. The concentration of fibres in water (consistency) would, at this stage, usually be between 0.1 and 1%. Clearly the rate at which the fibre suspension drains is an important factor in influencing the sheet formation process. Because fibres which have been heavily refined have a high surface area and are swollen, they drain slowly, whereas unrefined fibres drain very quickly. Many types of de-watering aids are used to assist the drainage process but these are essentially mechanical and it is beyond the scope of this book to discuss them in detail.

Models have been developed for this drainage process which are based upon theories of filtration. The Kozeny–Karmen equation is the most common rate expression used as a model for this filtration process. It can be expressed as:

$$\frac{\mathrm{d}Q}{\mathrm{d}t} = \frac{1}{K} \frac{(1 - C)^3}{S^2 C^2} \frac{1}{\mu} \Delta p$$

where $\mathrm{d}Q/\mathrm{d}t$ = rate of drainage per unit cross-sectional area of the web, $\Delta p$ = pressure gradient across the web, $C$ = volume fraction of the web occupied by solids, $S$ = specific surface area of the solids per unit volume and $\mu$ = viscosity of the liquid.

There are a number of weaknesses in this approach, amongst

which are that the flow may not be laminar in the case of medium to fast machines, and that the mat is also not incompressible.

The wet web after the formation process is typically around 85% moisture and yet it is a characteristic of such networks that they have enough physical strength to be transferred at high speeds to the pressing section of the machine. During pressing, the sheet structure is consolidated by the physical expulsion of water, and the moisture content is reduced to around 65%. This is then followed by drying over steam-heated drying cylinders to produce the final sheet. There is a great deal of technology involved in these processes but little chemistry and the subject is not addressed further in this text.

*Chapter 6*

# The Surface Chemistry of Paper and the Paper-Making System

## INTRODUCTION

The surface chemistry of the paper-making suspension is very important to the properties of the final sheet of paper and also to the smooth running of the paper-making process itself. The aqueous suspension of fibres and fillers and also the added chemicals need to be retained efficiently during sheet formation, and this retention process is controlled to a large extent by the surface characteristics of the individual components and by the molecular and colloidal interactions taking place in the aqueous phase. In addition, the effective functioning of many of the chemicals which are added depends upon their adsorption, conformation and orientation at the fibre or pigment surface. The following chapter attempts to cover some of the more important surface chemical aspects of the paper-making system.

## SURFACE CHEMISTRY OF FIBRES AND FILLERS

### Surface Chemistry of Fibres

The surface character of fibres influences their affinity towards various chemical additives by, for example, their adsorption properties and also in their tendency to flocculate. Cellulosic fibres, because of the presence of acidic groups which are introduced during chemical pulping and bleaching, are mildly anionic. These acidic groups may be carboxylic (COOH) or in some cases sulfonic acid ($SO_3H$) and they are able to dissociate to leave a net negative charge on the fibre surface as the solvated acidic proton is released. Carboxylic acid groups in cellulose arise from a number of sources: they are introduced during alkaline degradation which produces carboxylic acid end groups on the reducing end of the cellulose and hemicellulose chains (see Chapter 3), they are produced in stopping reactions which stabilise the chain end to further degradation, they are introduced in oxidative treatments (*e.g.* bleaching), and they may be naturally present in hemicelluloses (*e.g.* the 4-*O*-methyl-D-glucuronic acid substituent groups in the glucuronoxylan of hard-woods). Sulfonic acid groups are introduced into the lignin of sulfite pulps and also sometimes into mechanical pulps by sulfite impregnation of the chips. The acidic group content of pulps may be expressed as an ion exchange capacity, and typical values for Kraft and sulfite pulps are shown in Figure 6.1.

As pulping progresses, the lignin content decreases and the acid group content also decreases. Lignin is not usually measured directly but by the degree of oxidisability of the pulp using, for example,

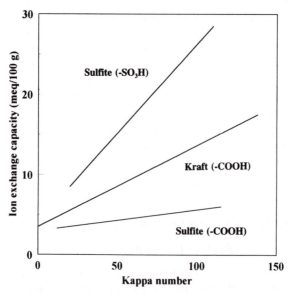

**Figure 6.1** *The acid group content of pulps expressed as an ion exchange capacity as a function of kappa number (measure of lignin content) for Kraft and sulfite pulps.*
(Source: 'Paper Chemistry—An Introduction', ed. D. Eklund and T. Lindstrom. DT Paper Science Publications, Grankulla, Finland, 1991, p. 20).

acidified permanganate. Kraft pulps have a higher carboxy content than sulfite pulps because they have undergone high-temperature alkaline degradation, but sulfite pulps contain higher levels of sulfonic acid groups. Bleached pulps also have a lower acid content than unbleached pulps, primarily because lignin and hemicelluloses are dissolved during the bleaching process. The anionic nature of the fibres manifests itself as a negative surface potential which can be demonstrated by measuring either streaming potential or micro-electrophoretic mobility (Figure 6.2). As would be expected, the surface charge is a function of the carboxyl content of the pulp (Figure 6.3).

The surface charge of a fibre has an important influence on its interaction with chemicals (both particulate and soluble) which are added to the aqueous fibre suspension. Their anionicity gives them a high affinity towards cationic additives, and many additives are produced in a cationic form in order to maximise their retention. The pH of the aqueous paper-making system is also important in these interactions. The surface charge of cellulose, because it arises from the dissociation of acidic groups, is dependent upon pH (see

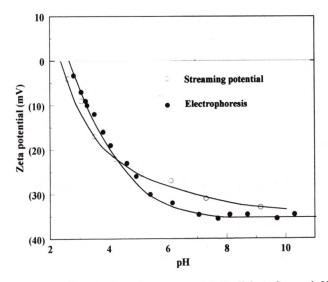

**Figure 6.2** *The effect of pH on the zeta potential of cellulosic fines and fibres as measured by streaming potential and microelectrophoresis (figures in brackets are negative).*
(Source: M.J. Jaycock and J.L. Pearson, *J. Coll. Interface Sci.*, 1976, **55**(1), 181 and *Svensk Papperstidning*, 1975, **5**, 167).

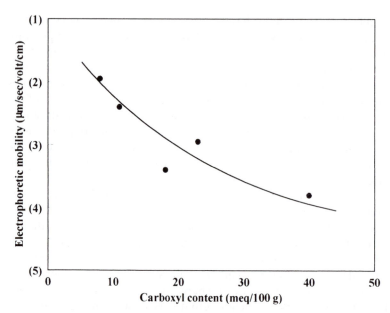

**Figure 6.3**   *Effect of carboxy group content on the microelectrophoretic mobility of microcrystalline cellulose (figures in brackets are negative).*
(Source: L.J. Sandell and P. Luner, *J. Appl. Polymer Sci.*, 1974, **18**, 2075).

Figure 6.2). At high pH the acidic groups exist in their dissociated salt form and the surface charge is substantially negative, whereas at low pH the acid groups exist in a largely undissociated form and the surface charge is closer to zero (the isoelectric point) or may even become positive. However, paper-making is not usually carried out at acidities low enough for the isoelectric point to be reached.

### The Surface Chemistry of Fillers and Pigments

Many inert pigments (often known as fillers) are incorporated into paper in addition to the cellulosic fibres. They may be added to improve certain optical properties—in particular opacity and brightness—or simply as a cheap replacement for costly fibre. The two most common pigments are kaolin (china clay) and chalk (limestone), but talc and speciality pigments such as titanium dioxide are also used. The particle size for general purpose fillers is normally expressed as an equivalent spherical diameter (esd) and this is determined from sedimentation data. Values for the common paper-making pigments are usually in the range 0.5–10 esd, but more specialised fillers (TiO$_2$, calcined clay, precipitated silica, *etc.*) often have more carefully controlled particle size distributions. The determination of esd assumes spherical geometry, but filler particles are not spherical and some consideration should therefore be given to the significance of this measurement. It is derived from sedimentation velocity measurements using Stokes' Law:

$$d = \left( \frac{18u\eta}{g(\rho - \rho_s)} \right)^{1/2}$$

where $d$ = the equivalent spherical diameter of the particle, $\eta$ = the viscosity of the solution, and $\rho$ and $\rho_s$ = the specific gravities of particle and solution respectively, $u$ = the sedimentation velocity, $g$ = acceleration due to gravity.

The problem with the measurement is that the relationship between $d$ and the actual particle dimensions is not clear, and the apparent value of $\rho$ for aggregated or porous particles is also difficult to obtain. It is also important to realise that, even if the particles were spherical, the relationship between actual diameter ($d$), surface area per unit mass ($S$) and particle numbers per unit mass ($N$) for particles of equal specific gravity would also not necessarily be linear (Table 6.1).

In practice, many fillers are either disc or rod-shaped and the relationship between actual diameter, surface area per unit mass and particle numbers per unit mass may be quite different from that of idealised spherical particles of the same esd and specific gravity (Table 6.2).

Particle sizes and surface areas for some common fillers which have been measured experimentally are shown in Table 6.3.

It is clear from Tables 6.1 to 6.3, that the effects of particle size

**Table 6.1** *Relationship between diameter, surface area and number of particles per unit mass for idealised spherical particles.*
(Source: R. Bown, 'Physical and Chemical Aspects of the Use of Fillers in Paper', in 'Paper Chemistry', ed. J.C. Roberts, Blackie, Glasgow, 1992, pp. 162–196).

| | | | | |
|---|---|---|---|---|
| Actual diameter | $d$ | $0.5d$ | $0.25d$ | $0.1d$ |
| Surface area per unit mass | $S$ | $2S$ | $4S$ | $10S$ |
| Particle numbers per unit mass | $N$ | $8N$ | $64N$ | $10^3N$ |

**Table 6.2**   *Theoretical multiplication factors for anisometric particles with an aspect ratio r ≫ 1. (Note: For a disc-shaped particle the aspect ratio is the ratio of disc diameter:disk thickness and for a rod-shaped particle it is the ratio of rod length:rod diameter).*
(Source: R. Bown, 'Physical and Chemical Aspects of the Use of Fillers in Paper', in 'Paper Chemistry', ed. J.C. Roberts, Blackie, Glasgow, 1992, pp. 162–196).

|                                                  | Spherical | Disc-shaped $r = 20$ | Rod-shaped $r = 20$ |
|--------------------------------------------------|-----------|----------------------|---------------------|
| Actual diameter (disc diameter or rod length)    | $d$       | $2.96d$              | $3.69d$             |
| Surface area per unit mass                       | $S$       | $2.48S$              | $3.70S$             |
| Particle numbers per unit mass                   | $N$       | $0.51N$              | $5.32N$             |

**Table 6.3**   *Particle size distributions and surface areas of some common fillers.*
(Source: R. Bown, 'Physical and Chemical Aspects of the Use of Fillers in Paper', in 'Paper Chemistry', ed. J.C. Roberts, Blackie, Glasgow, 1992, pp. 162–196).

| Filler | Particle size (mass %) | | | | $N_2$ BET surface area ($m^2/g$) |
|--------|---------|---------|---------|-----------|--------|
|        | > 10 $\mu$m | < 2 $\mu$m | < 1 $\mu$m | < 0.2 $\mu$m |        |
| Standard kaolin | 10 | 50 | 35 | 8 | 9 |
| Fine kaolin | 0 | 80 | 60 | 15 | 12 |
| Ground limestone | 1 | 60 | 35 | 10 | 7 |
| Precipitated calcium carbonate | 0 | 80 | 50 | 10 | 7 |
| Talc | 30 | 17 | 5 | 0 | 6 |
| TiO$_2$ (Anatase) | 0 | 96 | 94 | 8 | 23 |
| Silica | 6 | 40 | 30 | 18 | 150 |

and shape on surface area are of a similar order of magnitude. However, particle size is much more important than particle shape in influencing the numbers of particles per unit mass. Phenomena which are therefore dependent upon surface area will be influenced by shape and particle size, whereas those which are dependent upon particle numbers will be influenced more by size than shape.

Many fillers and pigments are cationic at acidic pH and thus can be deposited on to negatively charged cellulose fibres relatively easily. The isoelectric points of some common minerals used in paper making are shown in Table 6.4.

The charge or zeta ($\zeta$) potential of the filler particle (*i.e.* the charge at the plane of shear between the particle's diffuse double layer and the bulk liquid phase) can be obtained by measuring its mobility in an applied electric field of known magnitude. The mobility is a function of the field gradient and is therefore expressed as a speed per unit potential gradient ($\mu m/s/V/cm$). Mobility and therefore zeta potential are both a function of pH (Figure 6.4).

This means that the retention of fillers is likely to be very sensitive to pH and this is indeed found to be the case in commercial machine operation. There are also other important implications of the effect of pH on charge in the paper-making system, in particular on the interaction of fibre and filler components with cationic additives and these are discussed more fully both later in this chapter and in Chapter 7.

## Furnish Charge Measurement

Because of the importance of particle surface charge in chemical interactions of components of the aqueous fibre and filler suspension, paper makers would like to know the charge characteristics of all of the individual components of the aqueous suspension. However, such information is difficult to obtain experimentally, and some kind of average value is normally the best that can be hoped for in a multi-component system. The techniques used to determine furnish charge are usually one of those described below.

**Table 6.4** *Isoelectric points of some common fillers.*
(Source: R. Bown, 'Physical and Chemical Aspects of the Use of Fillers in Paper', in 'Paper Chemistry', ed. J.C. Roberts, Blackie, Glasgow, 1992, pp. 162–196).

| *Mineral* | *Isoelectric point* (pH) |
|---|---|
| $TiO_2$ (Anatase) | 6.0 |
| $TiO_2$ (Rutile) | 6.7 |
| Alumina | 9.3 |
| Silica | 2.0 |
| Kaolin | 2.0 |
| $CaCO_3$ (Calcite) | 8.3 |

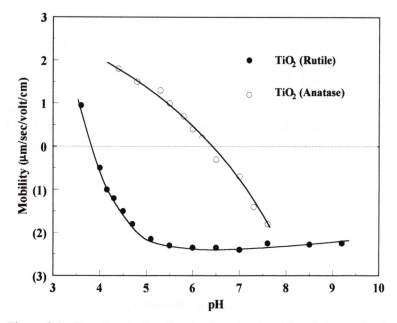

**Figure 6.4** *The effect of pH on the microelectrophoretic mobility of titanium dioxide particles (figures in brackets are negative).*
(Source: M.J. Jaycock, J.L. Pearson, R. Counter and F.W. Husband, *J. Appl. Chem. Biotechnol.*, 1975, **26**, 370).

'Microelectrophoresis (electrophoretic mobility)'. This involves the measurement of particle charge in an applied field. For paper furnishes, the supernatant solution—which contains finely divided colloidal matter, is usually removed and used to conduct the measurement. It must be questioned therefore as to how reflective this is of the charge characteristics of the larger particles and fibres which settle. However, as it is the colloidal fraction which requires to be flocculated to assist retention during drainage, it is still a useful measurement.

The microelectrophoretic mobility ($\mu_e$) is related to zeta potential ($\zeta$) *via* one of two equations. When the diameter of the particle is small relative to the thickness of the electrical double layer, the Huckel equation applies:

$$\mu_e = \frac{\zeta\varepsilon}{6\pi\eta}$$

and for particles whose diameter is large relative to the thickness of the double layer, the Smoluchowski equation applies:

$$\mu_e = \frac{\zeta \varepsilon}{4\pi\eta}$$

where $\varepsilon$ = dielectric constant and $\eta$ = viscosity.

'Streaming potential'. This is the second commonly used method and is often the basis of on-line measurements. It involves forcing a liquid through a capillary or a plug of porous material (*e.g.* pulp) by applying a pressure difference $\Delta p$. This causes a potential difference to be established across the plug (the streaming potential $V_s$). The streaming potential is related to zeta potential as follows:

$$\frac{V_s}{\Delta p} = \frac{\varepsilon \zeta}{\eta \lambda_o}$$

where $\varepsilon$ is the dielectric constant of the medium, $\eta$ is the viscosity and $\lambda_o$ is the electrical conductivity of the fluid.

Attempts have been made to use streaming potential for continuous on-line charge measurement. These methods involve the automatic formation of a pad of the furnish components through which white water (the recirculating drainage water produced during sheet formation) is passed. The streaming potential is then measured across the pad. These devices are used over a consistency range 0.2–0.6% and over a wide range of freeness (*i.e.* a measurement which is related to the speed of drainage of the suspension). The system then allows for regulation of the system zeta potential to what is considered to be an optimum level.

'Colloid (or polyelectrolyte) titration'. Charge may also be measured volumetrically using the principle of colloid titration which relies upon the fact that polymers of opposite charge can be stoichiometrically 'charge titrated' in aqueous solution. A cationic polyelectrolyte (C) can be titrated with an anionic polyelectrolyte in the presence of an appropriate indicator (I) as follows:

$$A + C \overset{k_1}{\rightleftharpoons} AC$$

$$A + I \overset{k_2}{\rightleftharpoons} AI$$

where A = an anionic polyelectrolyte, I = indicator, C = a cationic polyelectrolyte, AC = polyelectrolyte complex and AI = dye complex.

The anionic polyelectrolyte is usually potassium poylvinyl sulfate

(KPVS) and the indicator is usually $o$-toluidine blue (OTB). It is necessary for $k_1$ to be much larger than $k_2$ (which is usually the case for polyelectrolytes). If for example a cationic polyelectrolyte together with OTB is titrated with KPVS, a polyelectrolyte complex is initially formed until no free polyelectrolyte is left to react with the KPVS. At this point, KPVS starts to react with OTB and a colour shift from light-blue to purplish-red indicates the end point. The titration relies upon the formation of a 1:1 complex, which is generally true provided that the ionic strength is low.

The method can be applied to charges of solids such as fibres and fillers by equilibrating a known excess of cationic polyelectrolyte (for anionic paper furnishes) with the furnish. The solid phase is then separated and the residual polyelectrolyte in the filtrate is back-titrated:

$$\text{Fibre or filler} + \text{C} \rightarrow \text{Filtrate}$$

$$\text{Filtrate} + \text{A} \rightarrow \text{Complex}$$

From a knowledge of the stoichiometry of the reaction between A and C, the charge of the furnish can be determined.

## Dissolution from Pulps and Fibres

During refining, soluble substances are dissolved from pulps to the extent of about 2–5% by weight of the pulp, but in some cases this may be higher. This is not simply residual material left from incomplete washing (although there may be some) because it is found even in well-washed pulps, but it is the result of the mechanical forces of refining. The extracted material is usually a mixture of lignin-type material and hemicelluloses, the most dominant component of which is xylan. The components are polymeric (although of fairly low DP) and, because they are usually highly anionic, they compete for cationic polymers. They exert a 'cationic demand' for added cationic polyelectrolytes, and it is important to have some knowledge of their contribution to the charge of the system so that an appropriate choice of polymer treatment may be used. It is common to use highly charged cationic polyelectrolytes as a pretreatment to reduce the anionicity of the system. Some typical lignin structures identified in extracts from the recirculatory water from a newsprint machine are shown in Figure 6.5.

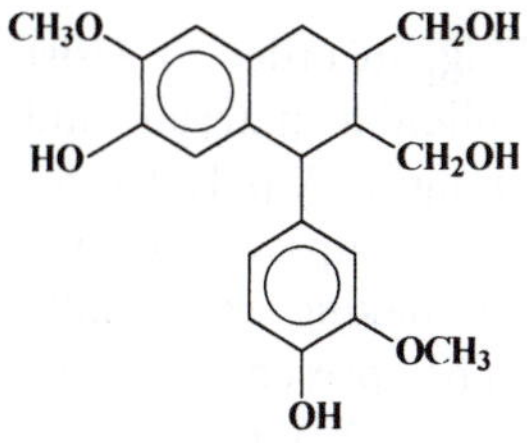


**Isolariciresinol**

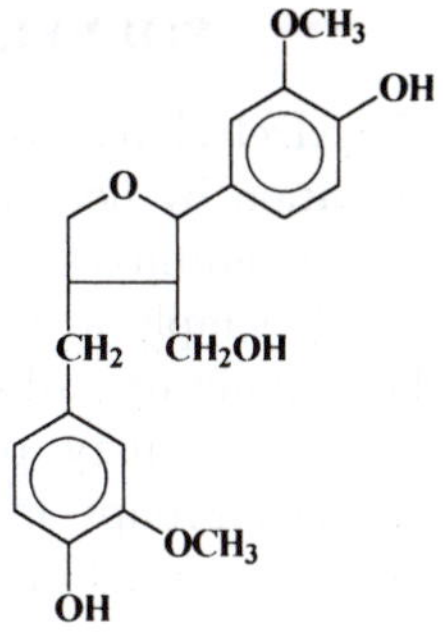


**Lariciresinol**

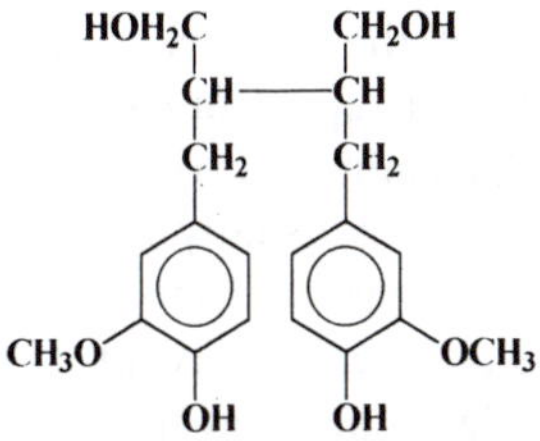


**Secoisolariciresinol**

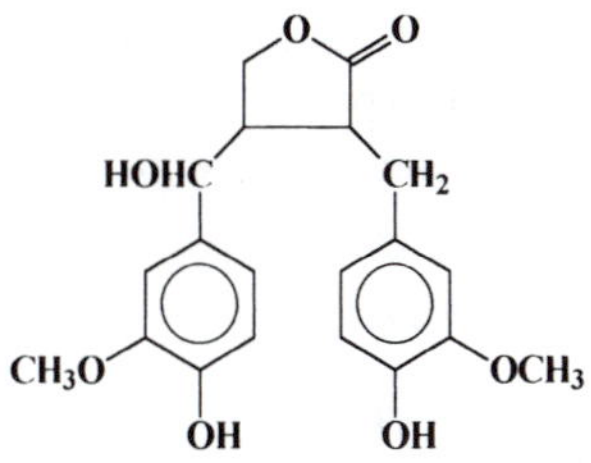


**Hydroxymatairesinol**

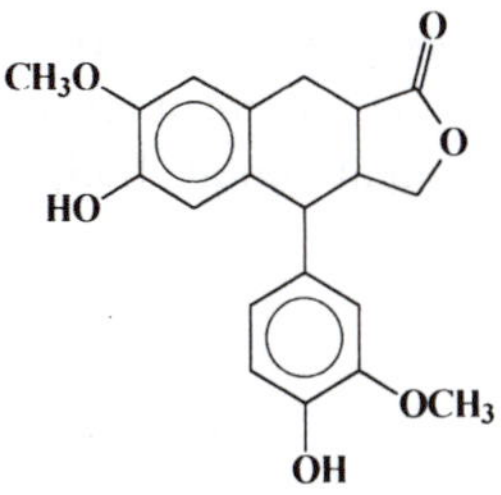


**α - Conidendrin**

**Figure 6.5** *Lignin structures identified in extracts from the white water from a newsprint machine.*
(Source: 'Paper Chemistry', eds. D. Eklund and T. Lindstrom, DT Paper Science Publications, Grankulla, Finland, 1991, p. 55).

## POLYELECTROLYTES IN PAPER MAKING

Polyelectrolytes are used widely in paper making, in order to assist retention by floccculation and aggregation of colloidal material, and also as additives for promoting wet and dry strength. A polyelectrolyte is simply a polymer which contains charged groups which may be anionic or cationic. The ones used in paper making are usually linear but may occasionally be branched. The presence of the charged groups makes the polymer water-soluble. A typical polyelectrolyte used in the process is shown in Figure 6.6.

### Characterisation of Polyelectrolytes

The way in which polyelectrolytes function is characterised primarily by two criteria, their molecular weight and their charge density. The charge density, which is also known as cationicity or anionicity depending upon the sign of the charge, is the ratio of the charged groups to the total number of repeat units in the polymer. This may not always be easy to determine for complex co-polymers with no clearly defined repeat unit. It is usually measured by polyelectrolyte (colloid) titration, which has been discussed earlier in this chapter, and the result expressed as a mole percent of charged groups or as equivalents or milliequivalents per unit mass. The cationic polyacrylamide shown in Figure 6.7 is typical of many used in paper making, and would have a mole percent cationicity of $m/(m + n)$.

### Adsorption of Polyelectrolytes

Polyelectrolytes are required to be adsorbed by fibres and fillers in order to perform their function, and the adsorption process is dependent upon both charge density and molecular weight. In

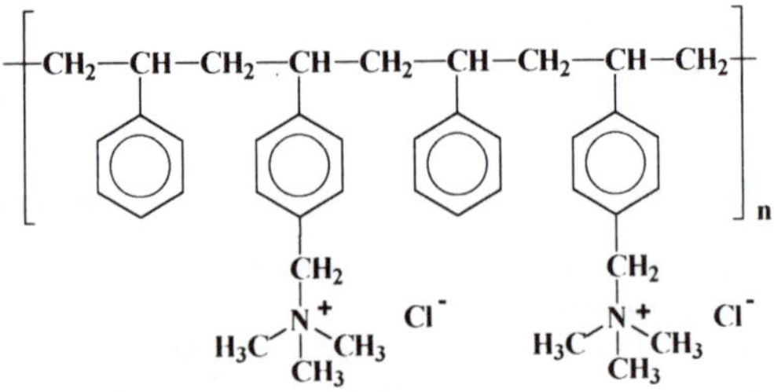

**Figure 6.6**  *A cationic polystyrene polyelectrolyte.*

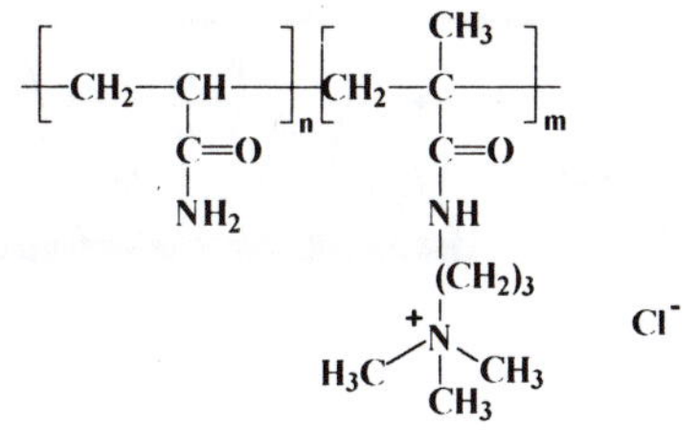

**Figure 6.7** *A typical cationic polyacrylamide used in paper making.*

general, a plot of adsorption against charge density for polyelectrolytes of different molecular weight would take the form shown in Figure 6.8.

Adsorption at high cationicity is low and relatively independent of molecular weight because the polyelectrolyte is adsorbed in a relatively flat conformation. Adsorption at low to intermediate cationicities is higher and also tends to be dependent upon molecular weight. This is because the polyelectrolyte is adsorbed in a much less compressed conformation. This is represented pictorially in Figure 6.9.

The adsorption of ionic polyelectrolytes by mineral fillers, fibres and fines is an essential first step in many chemical modification

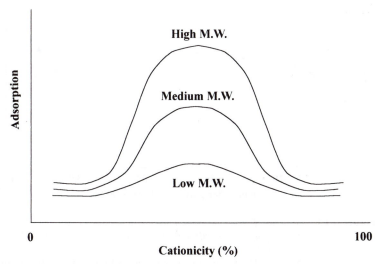

**Figure 6.8** *The generalised effect of molecular weight and cationicity on adsorption of cationic polyelectrolytes on non-porous surfaces.*

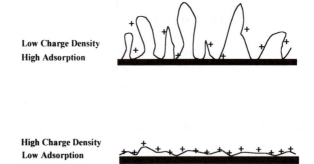

**Figure 6.9** *Pictorial representation of the conformation of adsorbed polyelectrolytes.*

processes, particularly in assisting their retention in the fibre web during sheet formation. The mechanism by which they assist retention is discussed in more detail in Chapter 7, but the adsorption process, as far as non-porous fillers is concerned, corresponds well to

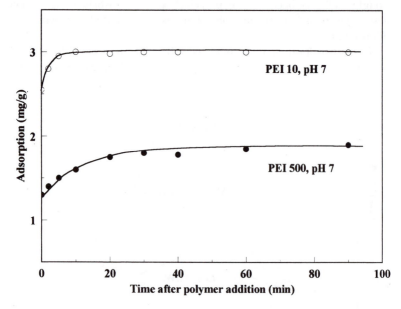

**Figure 6.10** *Influence of molecular weight on the adsorption of polyethyleneimines (PEI) by a bleached sulfite pulp (PEI 10 = DP of 10 and PEI 500 = DP 500).*
*(Source: D. Horn, in 'Polymeric Amines and Ammonium Salts', ed. E.J. Goethals, Pergamon Press, Oxford, 1980, p. 333).*

the above theoretical principles. However, the adsorption of poly-electrolytes by cellulose, because it is porous, is less simple and depends upon the pulp type, its accessibility and its porosity. These, in turn, are influenced by the conditions of stock preparation. Adsorption is relatively rapid and is usually fairly complete within a few seconds or minutes. An example for polyethyleneimines (PEI) is shown in Figure 6.10.

At comparable charge densities, the lower molecular weight PEI (PEI 10 = molar mass of 400) is adsorbed more easily than the high molecular weight PEI (PEI 500 = molar mass of 25 000) and this is the opposite of what is found for polyelectrolyte adsorption on to non-porous substrates (for example by pigments). The effect can be explained by the fact that cellulose, being porous, is less accessible to the high than the low molecular weight polymer.

The effect of charge density is, however, the same as for poly-electrolyte adsorption by non-porous substrates, *i.e.* the lower the charge density the higher the level of adsorption. For example, the

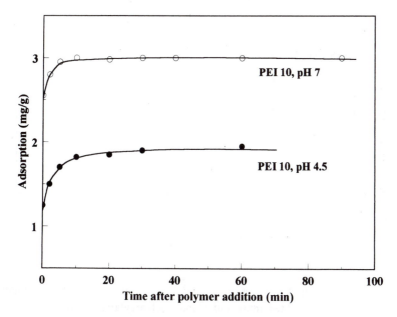

**Figure 6.11**   *Effect of pH on the adsorption of polyethyleneimine by a bleached sulfite pulp (PEI 10 = DP of 10).*
(Source: D. Horn, in 'Polymeric Amines and Ammonium Salts' ed. E.J. Goethals, Pergamon Press, Oxford, 1980, p. 333).

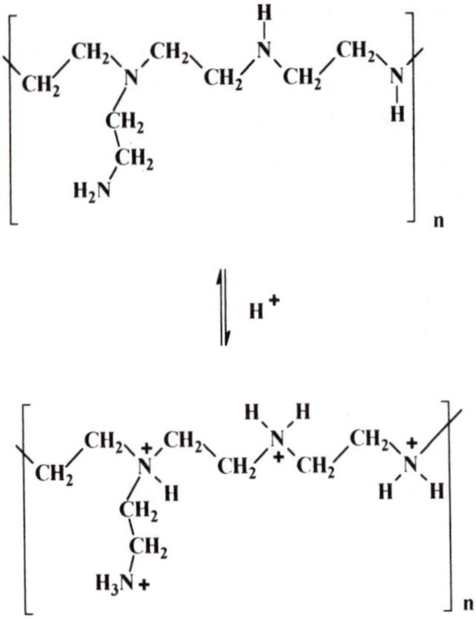

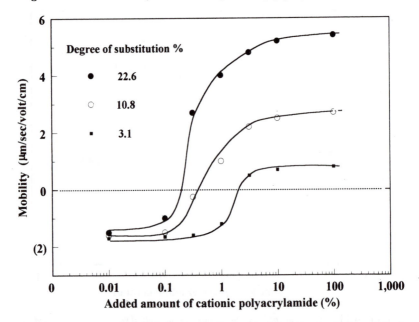

**Figure 6.12**   *Protonation of the amino nitrogen atom of polyethyleneimine.*

**Figure 6.13**   *The effect of charge density (degree of substitution of cationic groups) of a cationic polyacrylamide on the microelectrophoretic mobility of microcrystalline cellulose (figures in brackets are negative).*
(Source: T. Lindstrom, C. Soremark, C. Heinegard and S. Martin-Lof, *Tappi*, 1974, **57**(12), 94).

polyacrylamides tend to have a much lower charge density than the polyethyleneimines—typically about one fifteenth as much, but they are more highly adsorbed. The pH can also have an important effect upon charge density and therefore upon adsorption. For example, the polyethyleneimines are more strongly adsorbed at pH 7 than at pH 4.5 (Figure 6.11).

This is because their charge densities arise as a result of protonation of the amino nitrogen atom (Figure 6.12) and this increases at low pH. Since high charge densities give rise to lower levels of adsorption these polyelectrolytes are adsorbed less effectively at low pH.

The effect which polyelectrolyte adsorption has upon the surface charge (zeta potential) of fibres and fines is also important—particularly for retention—and both molecular weight and charge density of the adsorbed polyelectrolyte are known to affect the particle surface charge, although not always in an intuitively predictable way.

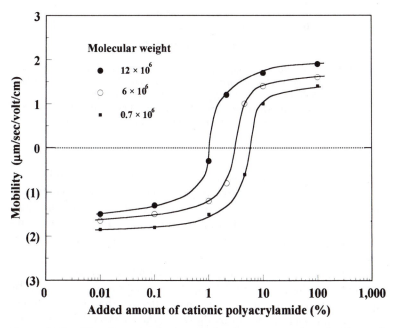

**Figure 6.14**   *The effect of molecular weight of a cationic polyacrylamide on the microelectrophoretic mobility of microcrystalline cellulose (figures in brackets are negative).*
(Source: T. Lindstrom, C. Soremark, C. Heinegard and S. Martin-Lof, *Tappi*, 1974, **57**(12), 94).

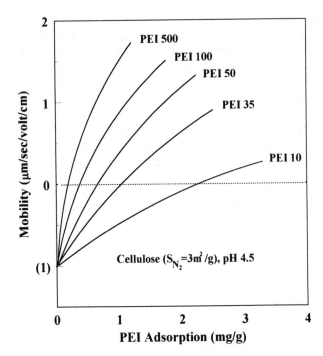

**Figure 6.15** *The effect of adsorption of polyethyleneimines of different molecular weight on electrophoretic mobility of cellulose (PEI 10 = DP of 10 and PEI 500 = DP of 500; figures in brackets are negative).*
(Source: D. Horn and J. Melzer, Trans. 6th Fundamental Research Symp., BPBIF, Oxford, 1978, pp. 135–148).

Firstly charge density; despite the fact that polyelectrolytes of comparable molecular weight are adsorbed in a flatter conformation and thus to a lesser extent as their charge density increases (see above), charge neutralisation is still achieved at a lower addition level for high than for low charge density polyelectrolytes (Figure 6.13).

The effect of molecular weight is also intuitively surprising. High molecular weight polyelectrolytes, although adsorbed less easily by porous substrates (see above), are more effective than lower molecular weight polyelectrolytes in modifying surface charge (Figure 6.14).

The phenomenon is also exhibited by the polyethyleneimines (Figure 6.15).

The reason for this is that the higher molecular weight polyelectrolytes tend to be adsorbed preferentially at the surface (as they cannot penetrate the porous structure so effectively) and are hence more effective electrokinetically at the liquid–particle interface.

The ionic strength of the solution also significantly influences polyelectrolyte adsorption. In general, the higher the ionic strength of the medium, the less extended and the more coiled the polymer conformation becomes (due to preferential interaction with counter ions in solution rather than with other segments of the polymer chain). The coiled polymer becomes more accessible to the internal porous structure and adsorption is increased. However, for the same reason, it is less influential on the surface charge.

### Reconformation

Polyelectrolytes, once adsorbed, undergo conformational changes which usually involve them adopting a more tightly bound conformation. This process causes the counter ion of the polyelectrolyte to

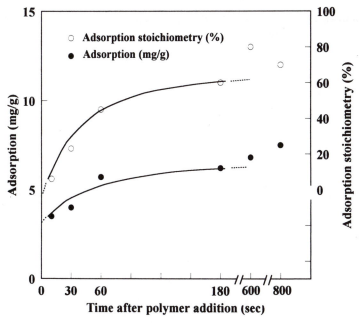

**Figure 6.16**  *Reconformation of adsorbed cationic polyacrylamide (MW $4 \times 10^6$) on cellulose fibres as shown by the kinetics of adsorption and adsorption stoichiometry (measured by counter ion release).*
(Source: L. Wagberg, L. Odberg, T. Lindstrom and R. Aksberg, *Colloids and Surfaces*, 1988, **31**, 119).

be released into solution instead of being tightly bound to the charge centre of the adsorbed polyelectrolyte. The measurement of counter ion release during and after adsorption can therefore be used to establish the degree of conformational change which the adsorbed polyelectrolyte undergoes. A high ratio of released counter ions (to total polyelectrolyte charges) indicates a relatively flat conformation as opposed to loops and tails. Reconformation of the polyelectrolyte at particle surfaces in the paper-making system has been demonstrated by this technique (Figures 6.16 and 6.17), although it is often a relatively slow process relative to the initial rapid adsorption.

**Initial Adsorption**                          **Reconformation**

**Figure 6.17**   *Pictorial representation of reconformation of an adsorbed cationic polyelectrolyte.*

*Chapter 7*

# Chemical Additives in the Paper Formation Process

## INTRODUCTION

Paper makers use many different synthetic and natural chemical additives for a variety of different reasons during the wet formation process. They are used to influence the efficiency of the formation process and also to impart specific sheet properties. They are usually added (with the exception of pigments) at a level of around 0–5% by weight of the other components of the furnish and, because of their relatively high cost, they often represent a significant proportion of the total raw material costs—particularly for recycled grades where the fibre costs may be very low. One of the most important areas of use is in the retention of fines and the dewatering and consolidation of the wet web. Additives used for this function are often polymeric and are frequently charged. They rely for their effect, as do many additives, upon adsorption to particle surfaces and for the influence that the adsorbed molecule has upon the state of flocculation of the dispersion.

Chemicals which are used to modify bulk sheet properties usually have to be added to the wet fibre suspension so that they become well distributed throughout the $z$-direction of the sheet. Chemicals which are added as a surface treatment to the dry sheet are usually only able to influence surface properties.

For the period from around 1840 to the early 1970s paper was usually made in an acidic environment at pHs of around 4–5. This was because many grades required the use of rosin and aluminium sulfate for the control of water penetration (sizing), and solutions of aluminium sulfate exhibit a pH of around 4.5. Aluminium sulfate has also been popular with paper makers because it assists the flocculation of colloidal particles and therefore behaves as a mildly effective retention aid. However, since the early 1970s there has been a move away from acidic systems towards neutral and even slightly alkaline pH. The advantages of operating at higher pH are that there is reduced corrosion, greater strength arising from better swelling of fibres at higher pH, the possibility of using high filler additions and the energy savings associated with the easier drying of filled paper. This change has had a profound effect upon the whole of the chemistry of the aqueous fibre suspension. The type of sizing system, retention aid and wet strength agent are all highly sensitive to pH. There is also a greater tendency for deposits, both chemical and microbiological, to develop in neutral or alkaline systems and this necessitates the increased use of biocides and pitch controlling chemicals. The shift to higher pH has also increased the use of dry strength agents, particularly cationic starch and polyacrylamides. This is because they help to offset the strength losses which arise from the use of higher levels of filler in paper.

This chapter attempts to describe some of the more important groups of chemical additives used in paper making and their mechanism of action.

## PAPER CHEMICAL USE IN SPECIFIC PRODUCT GRADES

*Newsprint*
This has historically involved only minimal use of chemical additives. The product is usually unsized and, because of its low cost, the use of polyelectrolytes for retention is not usually cost-effective. However, this situation is changing as a result of the trend towards limited filler inclusion and also the increased use of deinked waste paper. Newsprint is usually made in an acidic system due to the naturally low pH of groundwood pulp but, as more recycled fibre is used, the process may eventually become neutral.

*Printing and writing grades*
These may be coated or uncoated and these have been one of the strongest growth areas in the past few years. Much of this increase in demand can be accounted for by commercial printing and advertising (often light-weight coated), and by computer print-out and copier paper. Many of these printing and writing grades are now made under neutral or slightly alkaline conditions and in coating formulations there has been a move away from natural polymers and their derivatives (*e.g.* carboxymethyl cellulose and starch) towards synthetic polymers (*e.g.* acrylates, styrene–butadiene

co-polymers and polyvinyl alcohols and acetates). This is because of the better quality of these formulations for glossy advertising paper.

*Tissue and sanitary papers*
This has been a considerable growth area over the past twenty years. Because of the increased use of recycled fibre in tissue, deinking is of great importance. Softening agents are also used, particularly when recycled fibre, which produces less soft tissue, is used. Softening agents vary greatly in chemical type and mechanism of action. Debonding agents, in which a cationic water-soluble hydrophobic molecule is bound to the fibre and thus interferes with bonding, are popular but there are also those which change the textural quality of the sheet. The most common debonding agents in the former category are the quaternary ammonium type, with hydrophobic long-chain alkyl substituent groups.

*Packaging and board grades*
Many of these are made from recycled fibre, particularly in Europe. The chemical technology of recycling and deinking will therefore continue to be of importance in these products, and the control of pitch and deposits is therefore essential. Many grades require both wet and dry strength properties. There has been an increasing tendency to use alkaline systems and, hence, synthetic sizing for the production of many packaging grade boards, for example, aseptic liquid packaging board for such end-uses as juice cartons and fast food containers. Most liner board is also now made under alkaline conditions. Sizing characteristics are often very important in these products, and in particular the rate at which sizing develops is often a key requirement. In the future, new chemicals are likely to be developed as a result of the more stringent criteria which will probably be applied to chemical additives used in food packaging grades.

## RETENTION AND DRAINAGE AIDS

Retention and drainage aids are chemicals which are added to the fibre and filler suspension to assist the efficiency of the filtration process. Growth in recent years in the use of retention aids has been greater than that of almost any other paper chemical additive. It has been caused by a combination of factors: increased machine speeds, the increased use of filler in alkaline systems, the increased use of recycled paper and the growing tendency to use fillers in newsprint. Retention aids are water-soluble polymers which may be cationic, neutral or anionic. They are used either singly or in combination as flocculants, but the recent trend has been towards the increased use of cationic polyelectrolytes. The polyacrylamides and polyethylene-imines are the market leaders in this group but cationic starch in the presence of colloidal silica is becoming increasingly important. Many retention aids, in particular cationic starch and polyacrylamides also help to improve the dry strength of the paper. Retention is most easily defined with reference to a simplified wet formation system (Figure 7.1).

A low consistency suspension of fibres, pigments and chemical additives (thin stock) flows on to the moving wire mesh filtration medium of the paper machine where the wet web is formed. The water which drains away in this process (white water) is then recycled as far as it is practical to do so and is used to dilute the incoming high consistency suspension (thick stock). Retention can be considered either in overall terms or in terms of a single pass of the thin stock across the machine wire (first pass retention). For efficient operation, paper makers aim to achieve as high a first pass retention as possible, which they do by the use of retention aids. This reduces material losses and also assists in minimising the level of suspended solids in the effluent.

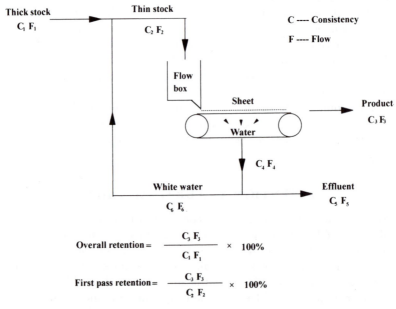

**Figure 7.1** *Simple schematic diagram of the sheet formation (wet-end) part of the paper machine.*

Retention aids can be classified into three types. First, 'inorganics', the most widely used of which are aluminium species such as aluminium sulfate and polyaluminium chloride. Some inorganics may also be used in conjunction with organic polymers, for example, silicic acid and cationic starch, or bentonite (montmorillonite clay) and nonionic polyacrylamides. The second important group is 'natural polymers' which are often modified and may be either charged (polyelectrolytes) or uncharged water-soluble polymers. Cationic starch is the most important of this group. Finally, 'synthetic polymers' represent probably the biggest group. These may also be charged or uncharged water-soluble polymers. Some typical classes of polyelectrolyte retention aid used in paper making are shown in Figure 7.2.

## Mechanisms of Retention

The mechanism by which retention is assisted is by some form of aggregation of the small particle size fraction of the fibre–filler suspension. This may involve similar particles (homocoagulation) or different particles (heterocoagulation). Polyelectrolytes of opposite charge to the surface of the dispersed phase are able to induce flocculation and therefore retention by primarily two types of mechanism: 'charge neutralisation' (including the charge patch model) and 'polymer bridging'. In the former case the initiation of flocculation *via* charge neutralisation and coagulation probably occurs by compression of the diffuse double layer at the particle surface and a reasonable interpretation can be made *via* classical DVLO coagulation theory. When two particles of like-charge approach each other, the potential energy of the interaction can be defined in terms of the sum of an electrostatic coulombic repulsive component and a short range van der Waals attractive component. Increases in ionic strength of the medium cause a compression of the electrical double layer and a tendency towards spontaneous flocculation (Figure 7.3).

This type of mechanism is likely to be partly operative in systems containing inorganic electrolytes as, for example, in the case of aluminium species. Some polyelectrolytes may also induce flocculation by charge neutralisation but the adsorbed polymer may also be able to bridge from one particle surface to another ('polymer bridging').

To determine which mechanism is operative, it is necessary to obtain information on the extent of adsorption of the polymer (adsorption isotherm), the effect of the adsorbed polymer on the

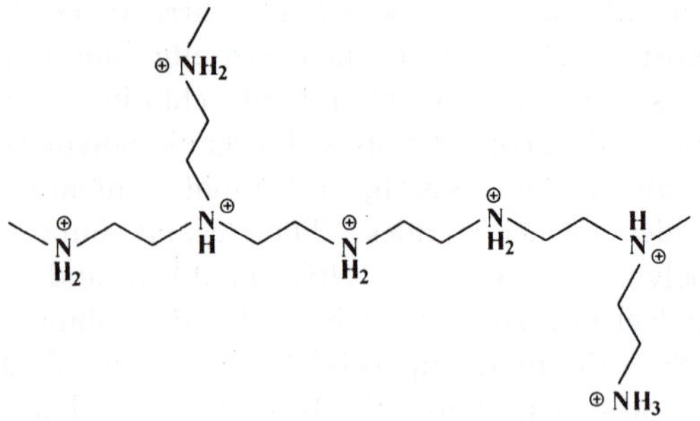

**Polyethylene imine**

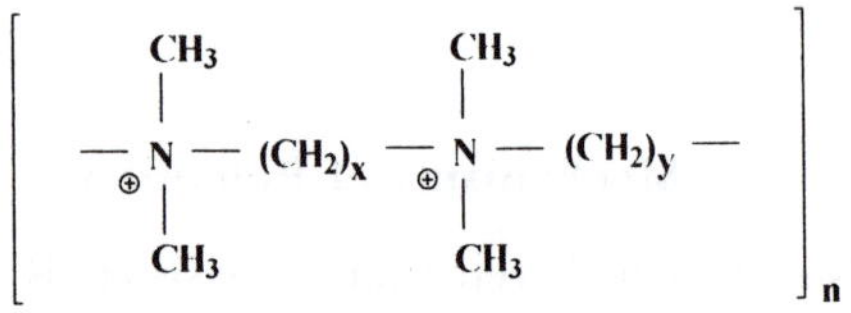

**x, y Ionene halides**

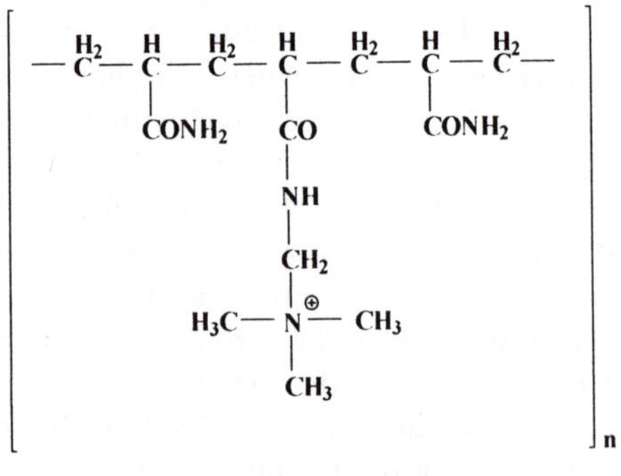

**Polyacrylamides**

**Figure 7.2** *Some typical cationic polyelectrolyte retention aids used in paper making.*

particle surface charge (*i.e.* mobility or zeta potential data), and to obtain some quantitative measurement of the extent of flocculation (*e.g.* the residual turbidity of settled suspensions).

'Charge neutralisation' is characterised by the isoelectric point (point of zero charge) being coincident with the point of optimum flocculation (minimum turbidity) — (Figure 7.4). A special case of

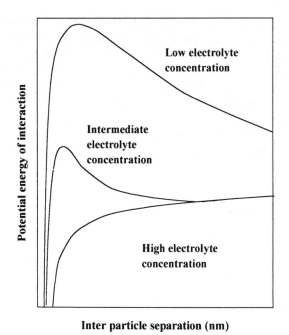

**Figure 7.3** *Effect of ionic strength on the energy of interaction between spherical particles.*

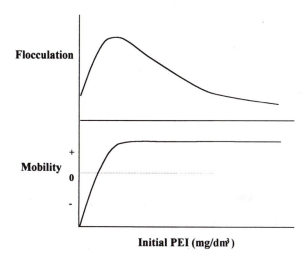

**Figure 7.4** *Typical flocculation and mobility behaviour for flocculation by charge neutralisation.*

charge neutralisation occurs when surface coverage is incomplete but the isoelectric point has been reached and flocculation has occurred (patch charge neutralisation—Figure 7.5). An explanation for this effect has been made in terms of what is known as the patch charge or charge mosaic model. In this model, flocculation is believed to occur by the attraction of patches of opposite charge, and it is shown schematically in Figure 7.6.

In some types of flocculation there is no coincidence of the isoelectric point with the point of optimum flocculation and 'polymer

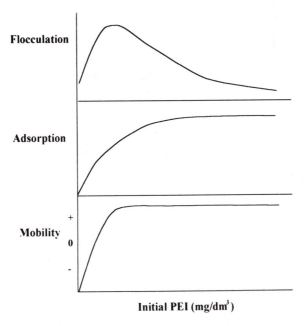

**Figure 7.5**   *Typical flocculation, mobility and adsorption data for flocculation by the patch charge neutralisation mechanism.*

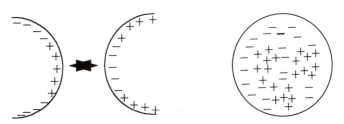

**Figure 7.6**   *Diagramatic representation of flocculation by the patch charge neutralisation mechanism.*

bridging', in which flocs are held together by polymer bridges, has been proposed as an explanation in these cases (Figure 7.7). Bridging has been demonstrated by direct experimental observation using neutron scattering.

High shear forces are prevelant in the approach flow system to the paper machine (*i.e.* as the fibre suspension approaches the point of deposition on the wire), and these have a large impact upon the efficiency of retention aids (Figure 7.8). A study of the effect of shear can often be helpful in establishing the mechanism of retention. Bridging flocculation is irreversibly sensitive to shear (*i.e.* when the shear forces are removed the suspension does not reflocculate) whereas charge neutralisation is reversibly sensitive to shear.

## DRY STRENGTH ADDITIVES

It is often necessary to increase tensile and other strength properties in paper by chemical means rather than mechanically *via* the refining process. This is usually the case when the other effects of refining, such as the decreases in opacity and air permeability, are not desired.

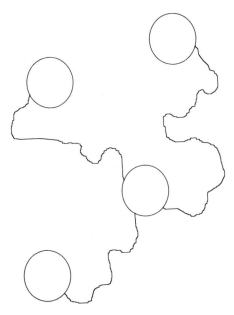

**Figure 7.7** *Diagramatic representation of flocculation by polymer bridging.*

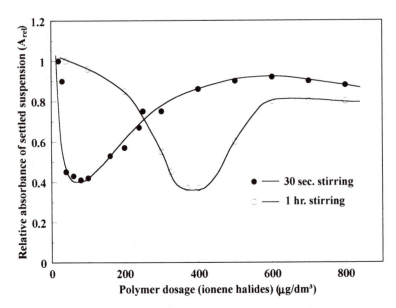

**Figure 7.8**  *The effect of shear forces (pre-stirring for different times) on the optimum polymer dose for ionene halide retention aids.*
(Source: Adapted from: J.W.S. Gosens and P. Luner, *Tappi*, 1976, **59**, 2, 89–94).

Dry strength additives are usually water soluble, hydrophilic natural or synthetic polymers, the commercially most important of which are starch, natural vegetable gums and polyacrylamides. These polymers are often made in cationic form by the introduction of tertiary or quaternary amino groups into the polymer, and are therefore polyelectrolytes. They are thus also able to function to some extent as drainage and retention aids.

Unmodified and anionically modified starches, soluble cellulose derivatives such as carboxymethylcellulose, polyvinyl alcohol, latex and other polymers are also used in some specialised applications. Starch, however, because of its cost, is by far the most common dry strength additive, about twenty times more being used than, for example, polyacrylamide.

Dry strength is an inherent structural property of a paper sheet which is due primarily to the development of fibre-to-fibre bonds during consolidation and drying of the fibre network. If we consider a sheet of paper failing under tensile load (Figure 7.9), it is clear that some fibres fracture and some are pulled out intact during failure. If the proportion of fibres broken is plotted against strength

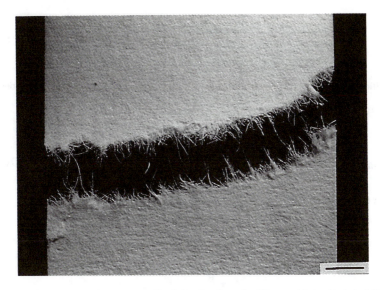

**Figure 7.9**    *Light photomacrograph of fracture line of a 15 mm wide tensile test strip of copier paper (60% hardwood, 40% softwood). Scale bar = 2 mm.*

for sheets refined to different extents, a broadly linear relationship is obtained (see Figure 4.5 in Chapter 4) and it is clear from this that the fibre-to-fibre bonds are weaker than the strength of individual fibres. The function of paper strength additives is therefore to assist bonding without adversely affecting the strength of single fibres.

### Starch

Starch is the most widely used dry strength additive and is normally made in a cationic form by introducing a reactive monomeric or polymeric tertiary amine or quaternary ammonium derivative into the molecule. The most commonly used reagent for tertiary amino starch is 2-chloroethyldiethylammonium chloride, and for quaternary starch is 2,3-epoxypropyltrimethylammonium chloride (Figure 7.10).

Tertiary amines require protonation to become charged and have little cationicity at high pH, whereas the quaternary starches are positively charged over a wide pH range (Figure 7.11). Cationic starches usually contain between 0.1 and 0.4% nitrogen which is equivalent to a degree of substitution of 0.01–0.05 (*i.e.* 1–5 charged glucose units per 100 in the polymer chain). They are usually applied in paper furnishes at a level of 0.25–2.50% based on fibre

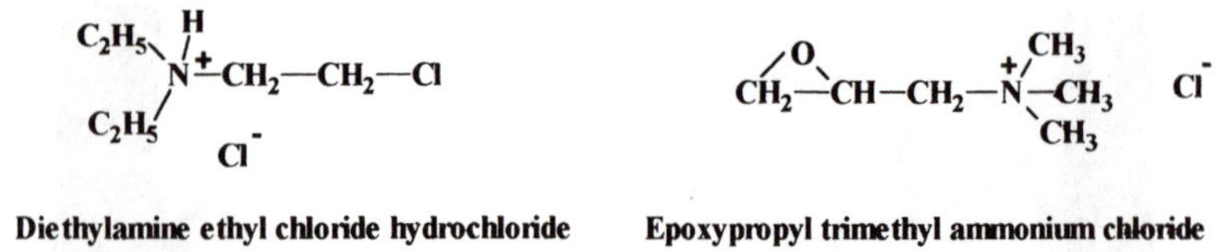

Diethylamine ethyl chloride hydrochloride          Epoxypropyl trimethyl ammonium chloride

**Figure 7.10**   *Common reagents for cationising starch.*

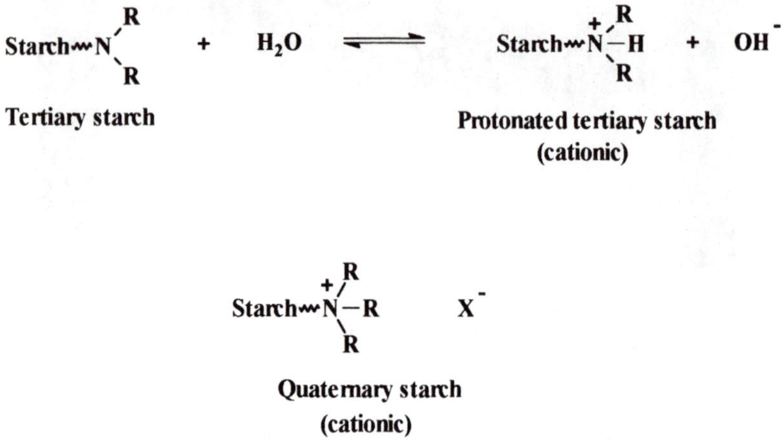

Tertiary starch                                          Protonated tertiary starch
                                                                (cationic)

Quaternary starch
(cationic)

**Figure 7.11**   *Protonation of tertiary amino cationic starches in aqueous solution.*

and are adsorbed quickly and virtually irreversibly. They are also retained more efficiently by unrefined fibres then refined fibres, because the starch has a high affinity for the high surface area fines present in the latter, and these tend to be preferentially lost during sheet formation (Figure 7.12).

Adsorption might be expected to create additional fibre-to-fibre bonds and/or to strengthen existing bonds, but there is evidence that the increase in tensile strength of paper is caused primarily by an increase in the bond strength per unit of optically bonded area, rather than by an increase in the extent of the relative bonded area itself (Figure 7.13).

The retention of cationic polymers by cellulose is most readily explained by ionic interaction between the cationic group of the polymer and the acidic groups of cellulose, but there is also evidence that hydrogen bonding participates to a lesser extent in the adsorption process.

Cationic starch is usually added to a blend of furnish components

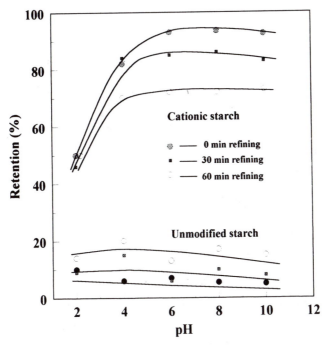

**Figure 7.12** *Retention of cationic and native starches during handsheet formation using fibres refined for 0, 30 and 60 minutes.*
(Source: J.C. Roberts, C.O. Au, K. Lough and G.A. Clay, *J. Pulp Paper Sci.*, Canada, 1987, 13, 1 J1–J5).

and is adsorbed preferentially by the high surface-area components. Thus, in a model furnish comprising 70% fibre, 15% pulp fines and 15% clay filler, it has been shown that the fibre fraction accounted for only one third of the adsorption of the applied starch, the remainder being almost equally adsorbed onto the fines and filler components. Large amounts of fillers and/or fines in a furnish therefore cause a reduction in the starch which is present on the fibre surface, and hence reduce the amount available to strengthen inter-fibre bonding. It follows that the order of mixing of the furnish components with starch, the addition sequences and the location of application points become important. For example, adding starch to fibres in the thick stock before filler addition favours filler retention, whereas adding starch to fillers separately, before fibre addition, or to the thin stock favours improvement of sheet strength. In the latter case, starch aggregates the filler and thus reduces its debonding effect at the price of some opacity loss.

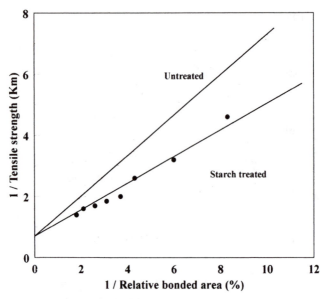

**Figure 7.13**  *Effect of cationic starch on the relationship between tensile strength and*
*relative bonded area.*
(Source: C.J. Jowsey, MSc Thesis, UMIST, Manchester,
England, 1986).

High shear conditions such as those which may exist in the fan
pump or at the pressure screens can also promote deposition of
charged high molecular weight polymers on to the long fibres.

## Vegetable Gums

Other natural polysaccharides of vegetable origin (gums) are also
used, but less frequently than starch. The most common of these are
the galactomannan polysaccharides derived from guar and locust
beans. The backbone chains in these gums are rigid, rod-like
mannose units with short galactose side-chains. Their molecular
weights are of the order of 200 000–300 000 daltons. Karaya gum is
a less frequently applied polysaccharide and has a somewhat more
complex side-chain structure, and a molecular weight close to 10
million daltons. They are very hydrophilic and form very viscous
aqueous solutions. They are not usually cationised and are probably
adsorbed through van der Waals forces and H-bonding to fibre
surfaces. The high surface area of the fines attracts a major portion
of the applied gums in a similar manner to the distribution of

cationic starch. There is some evidence that these gums may also act as colloidal protecting agents by reducing the rate of flocculation of long-fibred pulps and thus assisting the mass distribution (formation) of the sheet. The most potent deflocculation agent used for this purpose is the very high molecular weight deacetylated karaya gum which is used in the presence of alum. Guar gum also seems to reduce the friction of fibre suspensions in turbulent flow.

## Polyacrylamides (PAMs)

Polyacrylamides have the basic structure shown in Figure 7.14. Both anionic and cationic PAMs are used. Anionic PAMs have a poor affinity for fibres and are made fibre-substantive by using alum or a mediating cationic polymer. Cationic PAMs on the other hand are readily adsorbed by anionic fibres and furnish components without the need for alum. They can be used over a wide pH range, and also assist in drainage by acting as retention aids. Anionic PAM (0.25%) has been found to improve the tensile strength in a moderately beaten pulp by 18%, and 0.5% gave a 32% increase.

For strengthening effects, the molecular weight must be high enough to ensure effective adsorption and to provide multiple sites for H-bonding without causing bridging and flocculation of the fibres. The practical molecular weight range is usually between 100 000 and 500 000 daltons. Unlike refining, no large change in apparent density is observed when any of the PAMs are used for dry strength development, but tensile, burst, fold, internal bond strength and stiffness are all usually improved. They probably provide additional bonds between fibre surfaces where the distance is too great for H-bond formation between adjacent OH groups on cellulose. These new bonds are long and flexible enough to work without establishing optical contact between the bonding sites. The fact that they are able to double strength with very little change in density, porosity or light scattering is the most convincing evidence for this hypothesis. Short-bridging from one surface to another also may explain why they are particularly effective in increasing plybond strength.

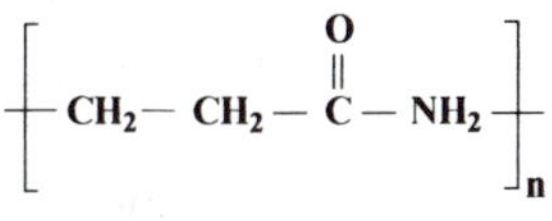

$$\left[ CH_2 - CH_2 - \overset{\overset{\displaystyle O}{\displaystyle \|}}{C} - NH_2 \right]_n$$

**Figure 7.14**  *Basic structure of a polyacrylamide.*

### Effect of Dry Strength Additives on Formation

Because dry strength additives are often polyelectrolytes, they are able to behave as retention aids. This leads to flocculation of the fibres and to mass distribution (formation) variations in the sheet which can reduce strength. The flocculation can be destroyed irreversibly by shear, which suggests that bridging flocculation is involved.

## INTERNAL SIZING

The term 'sizing', as used by paper makers, is somewhat ambiguous. It is used to refer both to the control of water penetration in the body of a sheet of paper or board (internal sizing) and also to the control of penetration through the surface of the sheet (surface sizing). The latter is strictly a surface modification and is therefore performed as a dry-end process. It is discussed more fully in Chapter 8. Internal sizing on the other hand, because it is intended to modify the water absorbing properties of the component fibres in the body of the sheet, is necessarily a wet-end operation. In this chapter, only the wet-end process of internal sizing will be considered. The reader is referred also to Chapter 4 in which the penetration of water and other fluids into paper is discussed more fully.

Internal sizing may be carried out over a wide pH range, but it is popular now to use neutral or slightly alkaline systems. When successfully performed, it retards the rate of penetration of a fluid through capillaries formed both within and between fibres. The fluid involved is, for most commercial grades of paper, water but resistance to non-aqueous fluids may also be important in some applications. This discussion will concern itself only with sizing against aqueous systems.

Internal sizing involves the modification of the fibre surface by a hydrophobic molecule so that the fibre surface energy is lowered. The sizes are normally added to the fibre–filler suspension as emulsified water-insoluble waxy materials with a particle size of around 1 $\mu$m. These must be retained in the wet web during its formation without interfering with fibre bonding and must then be able to migrate to non-bonded fibre surfaces during pressing and drying. The process therefore has three distinct phases:

'Retention' of the size emulsion particle in the wet web during sheet formation.

'Spreading' of the size molecules over the fibre surfaces.
'Orientation' at the fibre surface to create a low energy hydro-phobic surface.

Retention at the wet end is achieved by electrostatic interaction between a particle of the size emulsion, which is normally cationic, and the fibre surface. It is important that, at this stage, the size molecules do not spread over the surface and become orientated—as happens in the case of water-soluble quaternary ammonium salts with hydrophobic alkyl groups which are used as softening agents. This would have the effect of interfering with inter-fibre hydrogen bonding which is necessary for the creation of strength. In the second stage of the process, *i.e.* the spreading of the retained size over fibre surfaces, the physical properties of the size molecule are of crucial importance. The molecule must be sufficiently stable to resist hydrolysis and yet mobile enough, under the conditions of drying to cover fibre surfaces. Finally, for orientation to be achieved, some form of strong electrostatic or covalent interaction between the polar end of the sizing molecule and the fibre surface is required.

Sizing chemistry is critically linked to the pH of the paper-making system, and the oldest and most established sizing system involves the use of wood resin acids and aluminium sulfate. The presence of aluminium sulfate gives rise to an acidic pH ($\sim$4.5), because the aluminium ion in aqueous solution exists as a hexahydrated complex capable of dissociation according to the equilibria shown in equations 7.1–7.3. Further equilibria are also possible.

$$[Al(H_2O)_6]^{3+} = [Al(H_2O)_5OH]^{2+} + H^+ \qquad (7.1)$$

$$[Al(H_2O)_5OH]^{2+} = [Al(H_2O)_4(OH)_2]^+ + H^+ \qquad (7.2)$$

$$[Al(H_2O)_4(OH)_2]^+ = Al(H_2O)_3(OH)_3 + H^+ \qquad (7.3)$$

The trend towards increased calcium carbonate usage and there-fore neutral pH paper making has meant that there has been a steady decline in the use of rosin and alum and a concomitant increase in the use of sizes which are effective at higher pH. Commercially the most important in this group are the alkenyl succinic anhydrides (ASA) and the alkyl ketene dimers (AKD). These are able to esterify fibre surfaces directly and are more effective at neutral to high pH.

## Rosin and Alum at Acid pH

Rosin is obtained as an extract from softwoods. It is a mixture of a large number of closely related diterpene acids of which the most well-known is abietic acid (see Figure 2.9 in Chapter 2). They are used in either their free acid form (rosin acid) or as their sodium carboxylate salt (sodium rosinate). The latter is available as a fully or partially saponified product. The fully saponified form is water soluble but cannot be easily handled at consistencies greater than 50% whereas the partially saponified products are emulsions and can be handled at levels as high as 80% solids. The efficiency of rosin as a sizing agent can be improved by reacting the rosin acids with maleic anhydride *via* the Diels–Alder reaction (Figure 7.15).

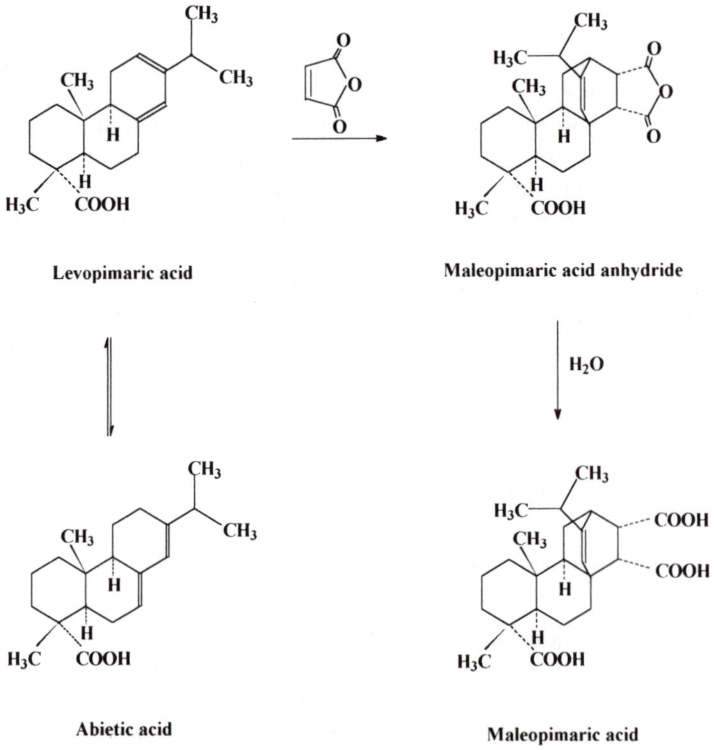

**Levopimaric acid**

**Maleopimaric acid anhydride**

$H_2O$

**Abietic acid**

**Maleopimaric acid**

**Figure 7.15** *Improvement of the efficiency of rosin as a sizing agent by Diels–Alder reaction with maleic anhydride.*
(Source: Adapted from J.M. Gess, in 'Paper Chemistry', ed. J.C. Roberts, Blackie, Glasgow, 1991, ch. 7, p. 181).

The product has two additional carboxy groups and is significantly more hydrophilic than the original starting material, yet it performs better as a sizing agent. This may be due to its increased anionicity which decreases its tendency to agglomerate and increases the ease and uniformity with which it spreads in the paper.

## The Interaction between Rosin and Cellulose Fibres leading to Sizing

There are various theories for the mechanism of sizing in a rosin–alum system. The first is that rosin in either its free acid or saponified form reacts with aluminium sulfate and the product provides the anchoring and orientation of the rosin molecule onto the surface of the fibre. There is evidence that different mechanisms are operative for saponified and unsaponified rosin. The reaction product of saponified rosin and alum is a positively charged product which is substantive to the anionic cellulose at the pH of the system. The unsaponified rosin on the other hand is probably precipitated onto the fibre in some way and then migrates through the sheet where it eventually reacts with alum to give a product analogous to that which is formed from the reaction of sodium rosinate and alum. It is uncertain how the rosin distributes itself through the paper, but melting and a physical flow or vapour phase migration have been proposed. The type of bond that is formed between rosin, the alum and the substrate is also uncertain, but no covalent bonding is thought to be involved. The mechanisms are complex but, at high pH, the size is either poorly retained or is retained in an inappropriately oriented form and this leads to complete loss of sizing. This has restricted the use of rosin–alum sizing to pH generally less than 6.

## Neutral pH Sizing Systems

Unlike the rosin–alum system, neutral sizing systems depend upon the size molecule undergoing covalent attachment to the hydroxy groups of the fibre surface through reactions such as esterification. The covalent linkage allows the relatively permanent attachment of hydrophobic groups in a highly oriented state which makes sizing possible at very low levels.

The main requirement of the molecule is that it has good stability towards water, since it is necessary to prepare it as a stabilised aqueous emulsion for use at the wet-end, and that it also has good reactivity towards cellulose. These are, to some extent, mutually exclusive, and a compromise is therefore sought. In addition, the selected molecule must have physical properties which allow diffusion and migration during drying and must be sufficiently chemically reactive at these temperatures to undergo reaction with cellulosic hydroxy groups. The most commercially successful of these sizes are the alkyl ketene dimers (AKD) and the alkenyl succinic anhydrides (ASA) and their structures and reactions towards cellulose and water are shown in Figure 7.16. These are at opposite ends of the spectrum of reactivity and hydrolytic stability. AKDs are rather unreactive towards cellulose, except at elevated temperatures, but have good hydrolytic stability. The reverse is true for ASAs.

AKDs are waxy, water-insoluble solids with melting points around 50 °C, and ASAs are viscous water-insoluble liquids at room temperature. It is necessary to prepare them as stabilised emulsions by dispersion in a cationic polymer (normally cationic starch). Small amounts of retention aid and surfactants may also be present. Particle size distributions are around 1 $\mu$m, and addition levels around 0.1% (of pure AKD or ASA) by weight of dry fibre. This is an order of magnitude lower than the amount of rosin used in rosin–alum sizing (1–2%). Emulsions of AKD are more hydrolytically stable than ASA, and the latter must be emulsified on-site and used within a few hours.

As in the case of rosin sizing, the first step is to retain the emulsified size particle in the wet web. The mechanism of retention is probably by heterocoagulation of the cationic size particles to the negatively charged fibre surface. The charge characteristics of the stabilising polymer become important as demonstrated by the effect of pH on the retention of AKD emulsion particles stabilised with a tertiary cationic starch (Figure 7.17).

The amount of reacted size required to give sizing correlates well with the BET-surface areas of the pulps used (Figure 7.18). Using these results together with surface balance measurements, a surface coverage of the order of 4% of a planar oriented monolayer can be calculated to be necessary to give this level of sizing.

Pulp type is also important and, in general, unbleached pulps are easier to size than bleached pulps. Pulps with a high cellulose content are extremely difficult to size and require as much as 10 times the amount of reacted AKD to produce similar levels of sizing to those of Kraft pulps. This effect is not easily understood and cannot be explained in terms of surface area. The presence of fillers also increases the amount of size required, and sizing efficiency has

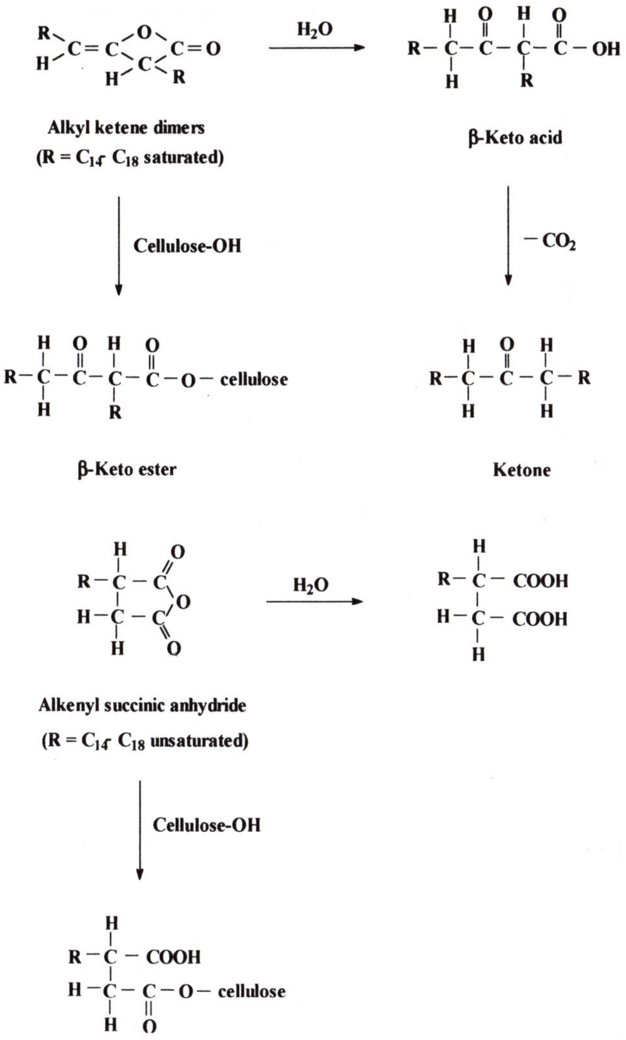

**Figure 7.16** *Reaction of AKD and ASA with cellulose and water.*

also been shown to be affected by the consistency, contact time, addition point and the presence of wet-end chemicals. The most important variables however are probably pH and drying conditions. For AKD sizing it is common for the sheet to be subjected to as high a temperature as possible to assist the reaction, and the final moisture content of the sheet may be as low as 1–3%. Sizing also

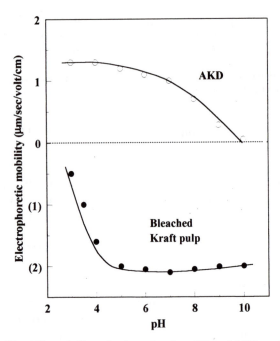

**Figure 7.17**   *Effect of pH on the electrophoretic mobility of AKD emulsion particles and of a bleached Kraft pulp.*
(Source: T. Lindstrom, G. Soderberg and H. O'Brien, *Nordic Pulp and Paper J.*, 1986, **1**, 26–42 and **2**, 31–45).

develops off-machine (Figure 7.19), and the uncertainty of the final level of sizing can be a serious problem when a narrow limit of sizing is a quality control criterion.

ASAs are prepared from alk-1-enes ($\alpha$-olefins) by catalytic iso-merisation, followed by an addition reaction with maleic anhydride (Figure 7.20). The location of the double bond in the alkene is important and it has been shown that the internal alkenes are much more effective than $\alpha$-olefins. The probable explanation for this is that the ASAs derived from $\alpha$-olefins, being solids at room tempera-ture, require higher temperatures for emulsification than the ASAs derived from isomerised olefins. The main advantage of ASA is that it is very reactive and, unlike AKD, no heat treatment is necessary and full sizing develops immediately off the paper machine.

The dicarboxylic acid product of ASA hydrolysis (see Figure 7.16) is inhibitory to sizing and this behaviour contrasts with that of AKD sizing, where the product of hydrolysis is not inhibitory. Both sizes

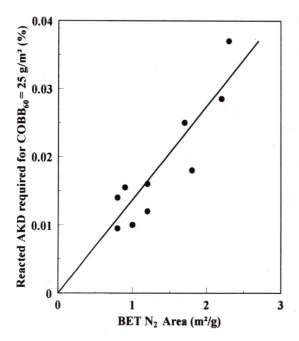

**Figure 7.18** *Correlation of reacted AKD size necessary to give sizing with the BET-surface areas of pulps.*
*(Source: T. Lindstrom, G. Soderberg and H. O'Brien, Nordic Pulp and Paper J., 1986, 1, 26–42 and 2, 31–45).*

also exhibit a tendency to lose sizing with time (reversion or fugitivity)—a phenomenon which is still rather poorly understood but which is of considerable commercial importance.

## WET STRENGTH

The tensile strength of paper and board falls by around 90–97% when it becomes wet (Figure 7.21), but many grades of product such as packaging papers, paper sacks, wall and poster paper, tea bag paper, *etc.* require a much higher retention of strength in the wet sheet. Wet strength agents therefore provide the sheet with the ability to retain a proportion of its dry strength when it becomes wet.

It is usual to express wet strength as a percentage of dry strength—usually tensile but not always; 15% is considered to be a minimum to qualify, although 20–40% is more common, and some

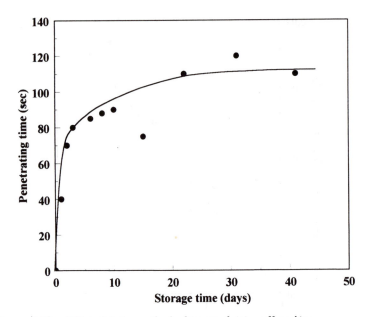

**Figure 7.19**  *Effect of drying on the development of sizing off-machine.*
(Source: J.C. Roberts and Z. Yajun, Proc. PIRA Symp. on
Neutral Paper Making, Stratford, England, 1990).

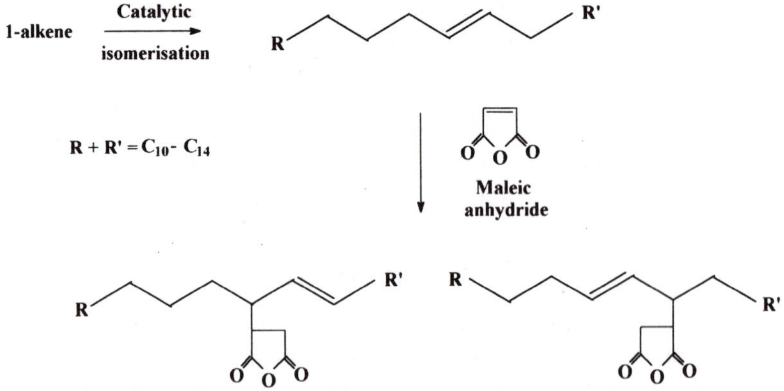

**Figure 7.20**  *The synthesis of alkenyl succinic anhydride (ASA)* via *the ene reaction
between maleic anhydride and catalytically isomerised alk-1-enes.*

papers may be as high as 50%. Sizing should not be confused with
wet strength, the former merely slows the rate of penetration of
water does not influence significantly the strength of the wet
sheet. However, it is common to use both sizing and wet strength

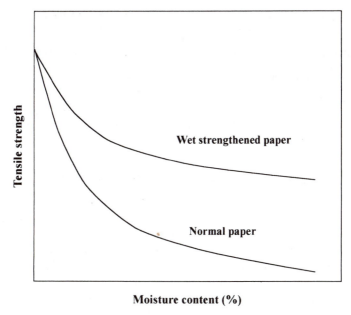

**Figure 7.21** *Diagrammatic representation of the tensile strength of normal paper and wet strengthened paper as a function of moisture content.*

agents in many grades. Wet strength is achieved by making inter-fibre bonds more permanent in the presence of water, and it is done by cross-linking.

### Wet Strength Additives

All wet strength agents are bi- or multi-functional molecules with the capability to cross-link with each other or with cellulose. The choice of chemistry depends to a large extent on pH. In acid systems, the main wet strength agents are urea–formaldehyde (U/F) and melamine–formaldehyde (M/F) resins, whereas in neutral and alkaline systems polyamine–polyamide–epichlorohydrin resins are more effective. However, these are not the only systems in use, and a summary of these and other available methods is provided in Figure 7.22.

Wet strength agents need to be dispersible in water, reactive and possess an affinity for the fibre so that they can be retained during sheet formation. This is usually achieved by the introduction of charged, often cationic, groups.

| | | |
|---|---|---|
| Urea–formaldehyde | pH | 3.8–4.5 (best at 4.0) |
| Melamine–formaldehyde | pH | 4.0–5.5 (best at 4.5) |
| Polyaminoamide–epichlorohydrin (PAE) | pH | 5.0–9.0 (best at 8.0) |
| Glyoxylated polyacrylamides | pH | 4.5–7.5 (best at 6.0–7.0) |
| Dialdehyde starch | pH | 4.5–6.5 |
| Polyethyleneimine (PEI) | pH | 7.0–9.0 |

**Figure 7.22** *Wet strength resins used commercially in paper.*
(Source: Adapted from N. Dunlop-Jones, in 'Paper Chemistry', ed. J.C. Roberts, Blackie, Glasgow, 1991, ch. 6, pp. 76–96).

## Mechanisms of Wet Strength Development

The precise mechanism of wet strength development is uncertain, but it is probable that it is brought about by either strengthening or protecting existing bonds or by forming additional water insensitive bonds. This may be done either by homocross-linking, in which the wet strength agent reacts with itself to produce a polymeric resin network which is physically entangled with the fibres and prevents rehydration and swelling, or it may be done by heterocross-linking in which the wet strength agent reacts with fibre components to produce inter-fibre bonds which are made permanent by cross-links.

## Urea–Formaldehyde Resins

Simply impregnating paper with formaldehyde and drying gives some limited wet strength, but it also causes brittleness and suffers from the problem of odour. The condensation product of formaldehyde and urea, 1,3-dihydroxymethylurea (Figure 7.23), is also effective, but it is water soluble and not substantive to cellulosic fibres in aqueous suspension.

However, in the presence of acid, 1,3-dihydroxymethylurea (dimethylolurea) condenses with itself to form low molecular weight resins (Figure 7.24). These are more effective, but are water insolu-

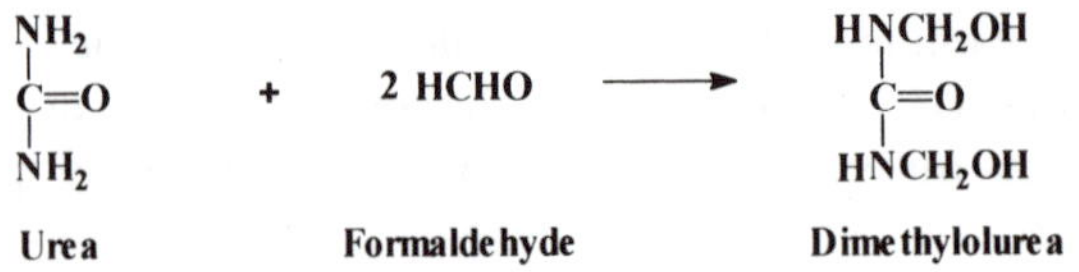

$$\begin{array}{ccc}
\text{NH}_2 & & \text{HNCH}_2\text{OH} \\
| & & | \\
\text{C}{=}\text{O} \quad + \quad 2\ \text{HCHO} \longrightarrow & \text{C}{=}\text{O} \\
| & & | \\
\text{NH}_2 & & \text{HNCH}_2\text{OH}
\end{array}$$

Urea              Formaldehyde              Dimethylolurea

**Figure 7.23** *Formation of 1,3-dihydroxymethylurea.*
(Source: Adapted from N. Dunlop-Jones, in 'Paper Chemistry', ed. J.C. Roberts, Blackie, Glasgow, 1991, ch. 6, pp. 76–96).

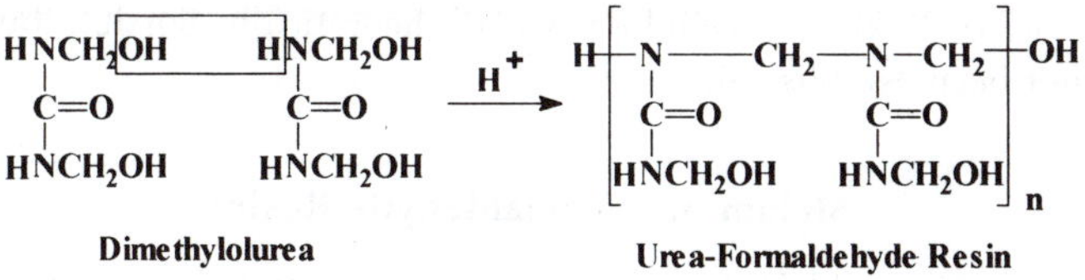

**Dimethylolurea**                    **Urea-Formaldehyde Resin**

**Figure 7.24** *Formation of urea–formaldehyde resins from 1,3-dihydroxymethylurea.*
(Source: Adapted from N. Dunlop-Jones, in 'Paper Chemistry', ed. J.C. Roberts, Blackie, Glasgow, 1991, ch. 6, pp. 76–96).

ble and can therefore only be used as surface treatments. They may, however, be made water soluble by introducing anionic $-SO_3H$ groups, and this also allows them to be made substantive to the fibres by precipitation on to the fibre surface with aluminium sulfate. They can also be made water soluble by the introduction of cationic amino groups. The products are normally used as aqueous solutions containing about 35–50% resin.

During drying of the paper, the relatively low molecular weight resins further polymerise to give a three-dimensional network which can protect existing bonds by homocross-linking (Figure 7.25).

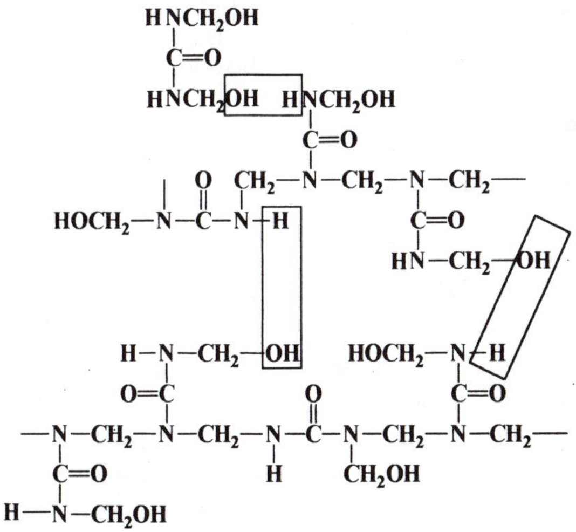

**Figure 7.25** *Further polymerisation of urea–formaldehyde resins.*
(Source: Adapted from N. Dunlop-Jones, in 'Paper Chemistry', ed. J.C. Roberts, Blackie, Glasgow, 1991, ch. 6, pp. 76–96).

Heterocross-linking to cellulose is also theoretically possible but this has not been established.

## Melamine–Formaldehyde Resins

Melamine–formaldehyde resins are also used and their chemistry is very similar to that of U/F resins (Figure 7.26). It is possible to react up to six moles of formaldehyde with melamine, but usually the trihydroxymethyl derivative is favoured. As these monomers are crystalline, two or more monomer units are usually condensed to give syrups. During drying of paper, high temperature and low pH promote cross-linking reactions by the formation of either ether or methylene linkages (Figure 7.27). As in the case of urea–formaldehyde resins, homocross-linking is the most probable mechanism.

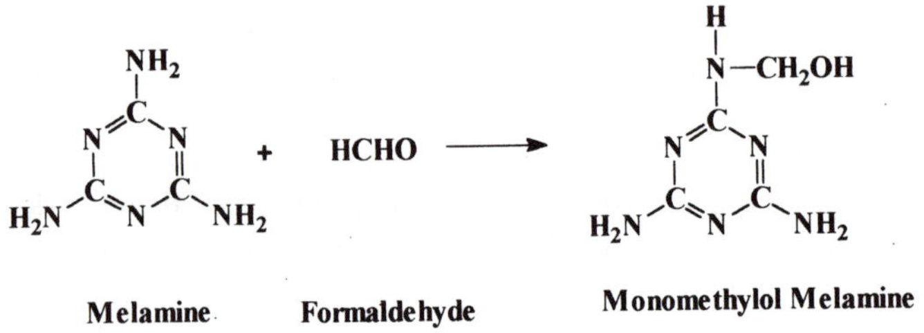

**Figure 7.26**   *Formation of monohydroxymethylmelamine.*
(Source: Adapted from N. Dunlop-Jones, in 'Paper Chemistry', ed. J.C. Roberts, Blackie, Glasgow, 1991, ch. 6, pp. 76–96).

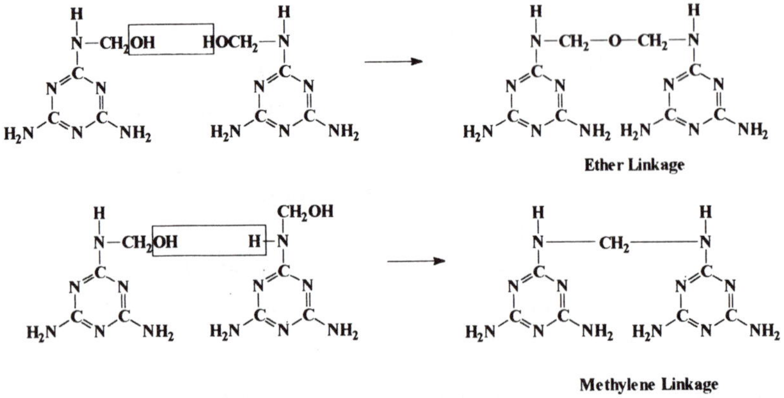

**Figure 7.27**   *Cross-linking reactions of hydroxymethylmelamines.*
(Source: Adapted from N. Dunlop-Jones, in 'Paper Chemistry', ed. J.C. Roberts, Blackie, Glasgow, 1991, ch. 6, pp. 76–96).

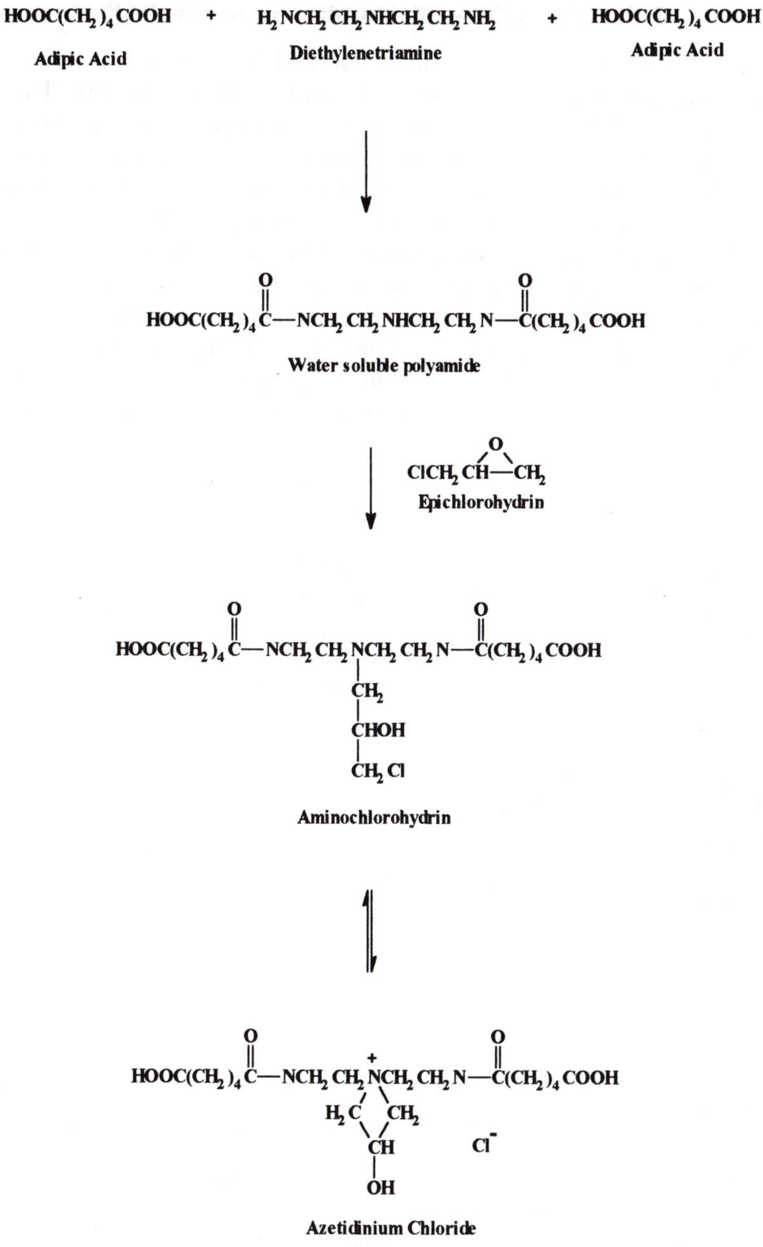

HOOC(CH$_2$)$_4$COOH   +   H$_2$NCH$_2$CH$_2$NHCH$_2$CH$_2$NH$_2$   +   HOOC(CH$_2$)$_4$COOH

**Adipic Acid**                    **Diethylenetriamine**                          **Adipic Acid**

$$\downarrow$$

$$\text{HOOC(CH}_2)_4\overset{\overset{\text{O}}{\|}}{\text{C}}\text{—NCH}_2\text{CH}_2\text{NHCH}_2\text{CH}_2\text{N—}\overset{\overset{\text{O}}{\|}}{\text{C}}\text{(CH}_2)_4\text{COOH}$$

**Water soluble polyamide**

$$\downarrow \quad \underset{\text{Epichlorohydrin}}{\text{ClCH}_2\overset{\overset{\text{O}}{\diagdown}}{\text{CH—}}\text{CH}_2}$$

$$\text{HOOC(CH}_2)_4\overset{\overset{\text{O}}{\|}}{\text{C}}\text{—NCH}_2\text{CH}_2\text{NCH}_2\text{CH}_2\text{N—}\overset{\overset{\text{O}}{\|}}{\text{C}}\text{(CH}_2)_4\text{COOH}$$

CH$_2$
CHOH
CH$_2$Cl

**Aminochlorohydrin**

$$\updownarrow$$

$$\text{HOOC(CH}_2)_4\overset{\overset{\text{O}}{\|}}{\text{C}}\text{—NCH}_2\text{CH}_2\overset{+}{\text{N}}\text{CH}_2\text{CH}_2\text{N—}\overset{\overset{\text{O}}{\|}}{\text{C}}\text{(CH}_2)_4\text{COOH}$$

H$_2$C   CH$_2$
CH        Cl$^-$
OH

**Azetidinium Chloride**

**Figure 7.28**   *Formation of polyaminoamide–epichlorohydrin (PAE) resins.*
(Source: Adapted from N. Dunlop-Jones, in 'Paper Chemistry', ed. J.C. Roberts, Blackie, Glasgow, 1991, ch. 6, pp. 76–96).

## Polyaminoamide–epichlorohydrin resins (PAE)

Unlike the urea– and melamine–formaldehyde resins, these wet strength agents are suitable for both neutral and alkaline pH. They are prepared by condensation of a dicarboxylic acid and bis(2-aminoethyl)amine, the free amino group is then alkylated with epichlorohydrin to give an aminochlorohydrin which exists in equilibrium with a 3-hydroxyazetidinium group (Figure 7.28).

The azetidinium group can theoretically cross-link to other amino groups by homocross-linking or to cellulose carboxy groups by hetero-cross-linking (Figure 7.29), but the former is more probable. Wet strength development is a function of curing temperature and time and it is also fairly permanent. Re-pulping is therefore difficult and this has implications for recycling these papers. Because of the

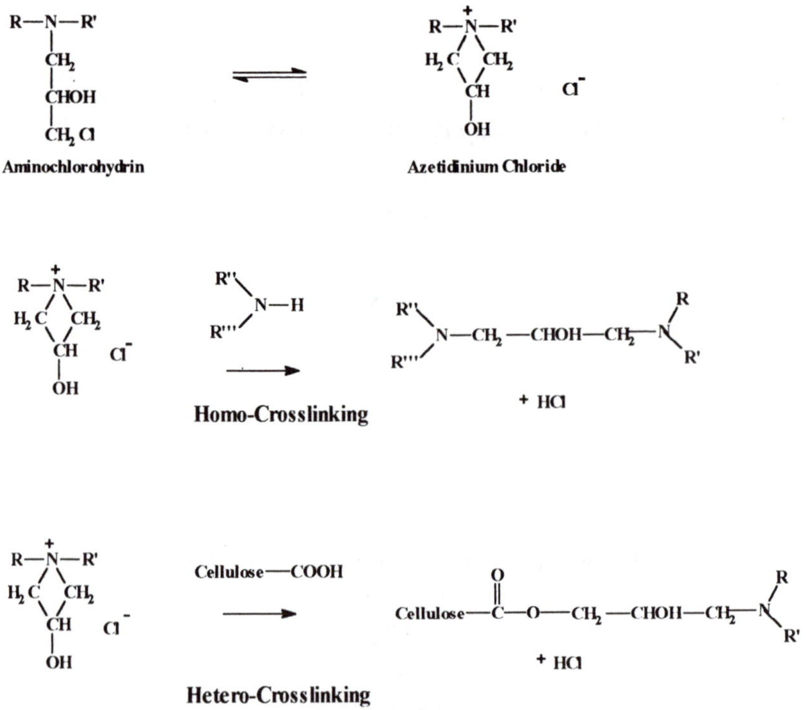

**Figure 7.29**   *Cross-linking reactions of PAE resins.*
(Source: Adapted from N. Dunlop-Jones, in 'Paper Chemistry', ed. J.C. Roberts, Blackie, Glasgow, 1991, ch. 6, pp. 76–96).

presence of charged groups, the PAE behaves as a polyelectrolyte and may also therefore affect retention.

## Glyoxylated Polyacrylamides

These are prepared from a polyacrylamide (which may be cationic) and glyoxal (Figure 7.30). The glyoxylated polyacrylamide may then participate in homo- or hetero-cross-linking (Figures 7.31 and 7.32).

In addition to these resins, polyethyleneimines and dialdehyde starches are also used to a lesser degree.

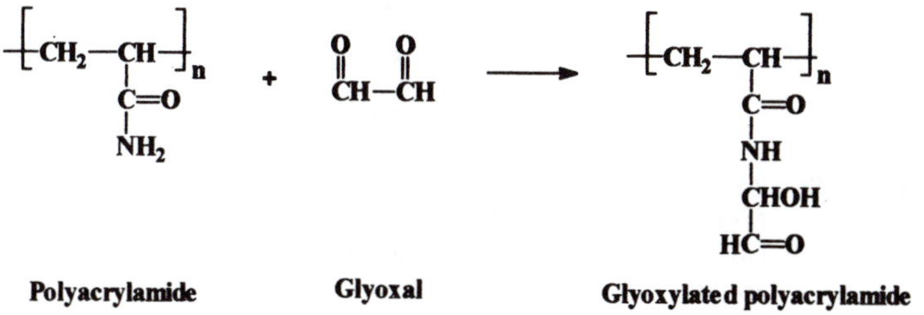

Polyacrylamide          Glyoxal          Glyoxylated polyacrylamide

**Figure 7.30** *Glyoxlation of polyacrylamides.*
(Source: Adapted from N. Dunlop-Jones, in 'Paper Chemistry', ed. J.C. Roberts, Blackie, Glasgow, 1991, ch. 6, pp. 76–96).

Polyacrylamide     Glyoxylated polyacrylamide

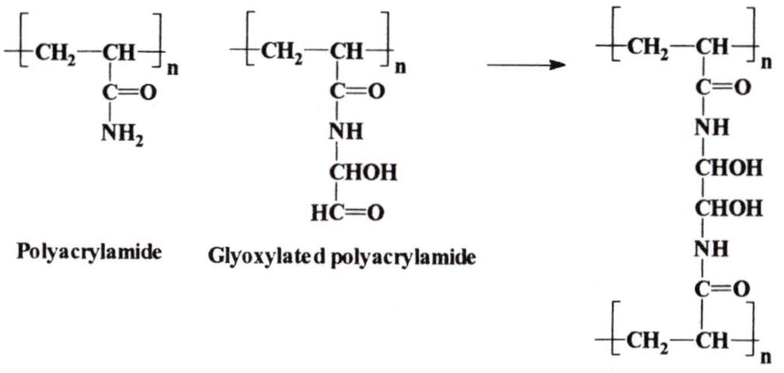

**Figure 7.31** *Homocross-linking of glyoxylated polyacrylamides.*
(Source: Adapted from N. Dunlop-Jones, in 'Paper Chemistry', ed. J.C. Roberts, Blackie, Glasgow, 1991, ch. 6, pp. 76–96).

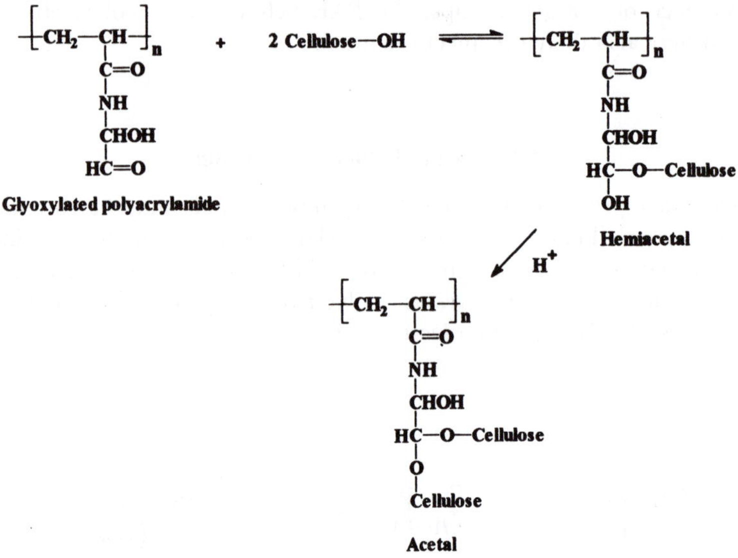

**Figure 7.32**  *Heterocross-linking of glyoxylated polyacrylamides with cellulose.*
(Source: Adapted from N. Dunlop-Jones, in 'Paper Chem-
istry', ed. J.C. Roberts, Blackie, Glasgow, 1991, ch. 6, pp.
76–96).

*Chapter 8*

# The Surface Modification of Paper

## INTRODUCTION

The surface of paper as it arrives directly off the paper machine is fairly rough and inhomogeneous and, for many applications—particularly those involving printing and writing where ink receptivity and image quality are important—it is necessary to improve its surface properties. This is done in one of two ways. The first of these is surface sizing which is carried out as an integral part of the paper machine operation (on-machine) after the sheet has been dried, and the second is coating which is usually carried out as a separate off-machine process. The surface sizing of paper involves the application of a water-soluble polymer (commonly starch or polyvinyl alcohol or a soluble cellulose derivative) to the surface of the sheet, and should not be confused with internal sizing (Chapter 7). The moisture which is picked up in this process is then removed by secondary dryers. The process has certain beneficial effects, particularly on printing characteristics. Coating, on the other hand, involves the application of a pigment-based coating mix to the surface of the paper, in a process which is radically different in both its effect and the technology of its application. Surface sizing provides a polymer film, through which the fibres of the base paper are fully visible, but which helps to consolidate the surface of the sheet and reduce its tendency to dislodge surface fibres (picking) during certain contact printing processes. Coating, on the other hand, provides a surface through which the component fibres of the base paper are largely invisible, and its main purpose is to produce a surface of high smoothness. This is demonstrated by the two photomicrographs in Figure 8.1.

This chapter attempts to describe the chemistry of these two processes in some detail.

(a)

(b)

**Figure 8.1** *Scanning electron photomicrograph of the surfaces of* (a) *surface sized and* (b) *coated paper. Scale bar* = 50 μm.

## SURFACE SIZING

Surface sizing is used to improve the surface quality of paper either as an end in itself or as a prerequisite to off-machine coating. When it is used as a pretreatment for coating its function is to improve the 'coating hold-out', that is the ratio, in volume terms, of coating remaining on the surface to that which penetrates into the base paper. This is affected by the viscosity of the coating mixture, the mean pore radius of the base sheet and the wettability of the base sheet as expressed by its contact angle. The wettability, and therefore contact angle of the base sheet, may be controlled by both internal sizing (see Chapter 7) and surface sizing.

### The Mechanics of the Size Press

The on-machine size press is the most common method of applying surface sizes to a base paper. The base paper is normally passed through two contacting rolls, the nip of which is flooded with the sizing solution. The paper absorbs part of the solution and the remainder is removed in the nip. The size press may be orientated vertically, horizontally or in an inclined configuration (Figure 8.2). The vertical configuration is the easiest to operate but gives rise to unequal uptake on either side of the sheet. This can be solved by the use of a horizontal configuration but the vertical approach of the paper makes this system difficult to operate. The inclined configuration is therefore often used as a compromise.

Ideally, the sheet should be uniformly covered with the sizing chemical but should not absorb so much water that excessive energy is needed in the after-drying operation. The hydrodynamics of the process are demonstrated in Figure 8.3. During the rotation of the rolls, the pond of sizing solution absorbs kinetic energy. Because there is an excess of liquid flowing towards the nip, and the nip pressure restricts the amount of solution which can pass through, the remaining solution flows back upwards in a circulatory motion. If the hydrodynamic forces are too large, this upward velocity causes

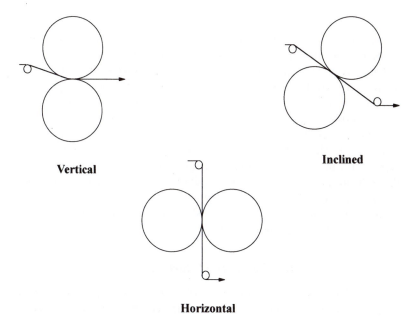

**Vertical**

**Inclined**

**Horizontal**

**Figure 8.2**  *Various configurations of the size press.*
(Source adapted from 'The Coating Processes', Tappi Press, 1993, p. 136).

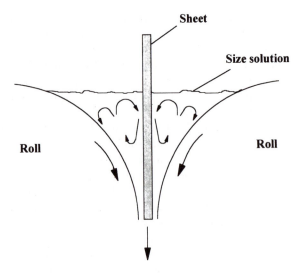

**Figure 8.3**  *Hydrodynamics of the size press.*
(Source adapted from 'The Coating Processes', Tappi Press, 1993, p. 136).

turbulence and an uneven pick-up of solids across the machine. This problem becomes more severe at high solution viscosities and, in order to avoid this turbulence at high machine speeds, it is necessary to use a relatively dilute size press solution, or to use a polymer of low molecular weight. The use of more dilute solutions, however, is less desirable because of the need for increased drying capacity.

A further problem of the size press operation is that of films splitting. At the exit of the nip, the film sometimes splits into two layers with one part following the paper and the other remaining on the roll surface. The problem manifests itself as small scale non-uniformities on the paper surface caused by threads of the size solution being drawn out and deposited unevenly on the surface to give a kind of 'orange peel' effect.

### The Chemistry of Surface Sizes

Surface sizes are usually solutions of water-soluble polymers. The most important of which, because of its commercial cheapness, is starch. Other more costly but more specialised film-forming polymers such as soluble cellulose derivatives (particularly carboxymethyl cellulose), polyvinyl alcohol and alginates are also used.

Starch is an energy storage polysaccharide which can be obtained in good yield from a number of sources such as corn, tapioca, potato, rice, *etc.* It exists as an insoluble granule of varying size depending on the source, and the granules are insoluble in cold water which allows for easy separation from the plant matrix by mechanical disruption, dispersion in water and sedimentation. The soluble polysaccharides within the granule are released by heating in water above the gelation temperature, at which point the granule swells and the component polysaccharides dissolve. The starch granule contains two starch polysaccharides, one is amylose, a linear $\alpha$-1,4-linked D-glucopyranose polymer, and the other is amylopectin, a branched polymer which differs from amylose in that it contains additional $\alpha$-1,6-branch points. The two structures are shown in Figure 8.4.

A summary of granular size, amylose and amylopectin contents and approximate degree of polymerisation of various starches is given in Table 8.1.

Native starches with a relatively high amylose content have a tendency to retrograde, a process which involves the irreversible precipitation of the amylose from the swollen gel. This is a problem when they are used on the size press and, whilst it can be avoided by using high amylopectin containing starches, these are less effective in terms of film formation. The tendency to retrogradation is controlled by chemical modification either by reducing its molecular weight which also allows a greater pick up because of the lower viscosity of the solution, or by introducing substituent groups which sterically hinder the molecular association process.

One of the difficulties of using starch as a surface size is that when surface-sized papers are recycled, the starch is readily solubilised into the aqueous system where it contributes to the nutritional pollution load of the effluent. This can be controlled to a large extent by introducing cationic groups into the starch molecule which bind strongly to the anionic cellulose in a similar manner to starches used as dry strength agents (Chapter 7).

Other water-soluble polymers which do not tend to suffer from problems of retrogradation are sometimes used when different properties are required. Soluble cellulose derivatives, particularly carboxymethyl cellulose, which is prepared by reaction of high purity cellulose with chloroacetic acid in the presence of alkali (equation 8.1), is popular for surface sizing base papers which are subsequently to be coated, because it assists in water removal when the coating mix is applied.

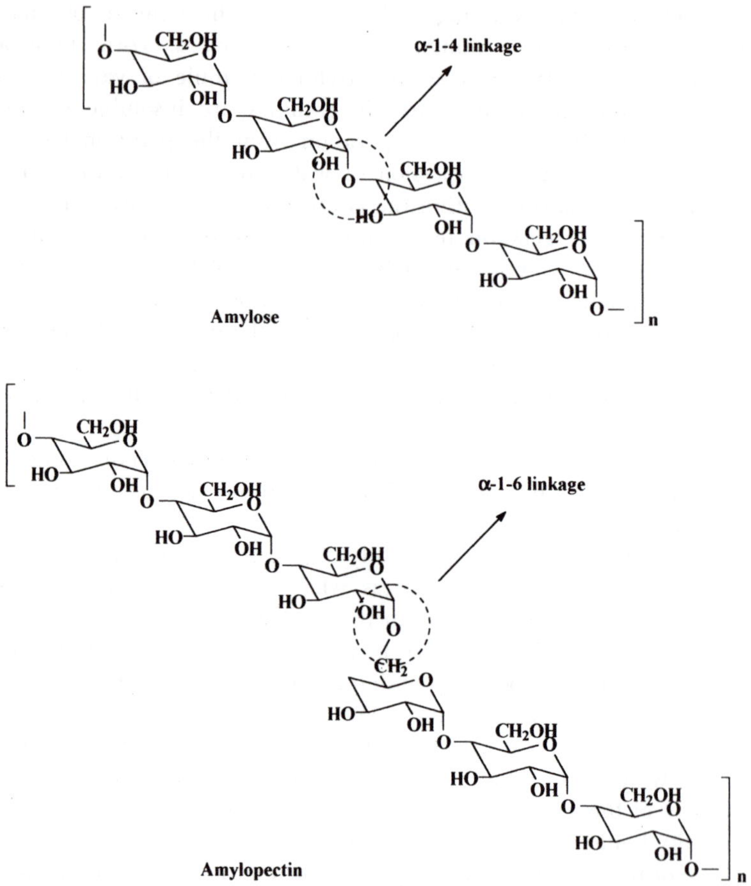

**Figure 8.4**   *Molecular structure of the amylose and amylopectin components of starch.*

Cellulose$-$OH + ClCH$_2$COOH + 2NaOH $\rightarrow$

$$\text{Cellulose}-\text{OCH}_2\text{COONa} + \text{NaCl} + 2\text{H}_2\text{O} \quad (8.1)$$

Carboxymethyl cellulose is normally used at a degree of substitution of around 0.1 (*i.e.* the number of hydroxy groups substituted per glucose residue), which renders the polymer water soluble. Further water solubility can be obtained by increasing the degree of substitution.

In grades of paper where a high density, clear, tough, well-sealed film on the surface is required, alginates may be used, sometimes in

**Table 8.1**   *The granular size, amylose and amylopectin contents of various starches.*
(Source: Adapted from R.L. Learney and H.W. Maurer, 'Starch and Starch Products in Paper Coating', Tappi Press, 1990, pp. 2 and 4).

| Starch | Granular size (μm) | Amylose (%) | Degree of polymerisation | Amylopectin (%) |
|---|---|---|---|---|
| Corn | 5–15 | 28 | 800 | 72 |
| Potato | 15–100 | 21 | 3000 | 79 |
| Wheat | 2–35 | 28 | 800 | 72 |
| Tapioca | 5–35 | 17 | 3000 | 83 |
| Waxy maize | 5–25 | 0 | | 100 |
| High-amylose corn | 5–40 | 40–70 | 600 | 30–60 |
| Grain sorghum | 5–30 | 25 | 800 | 75 |
| Waxy grain sorghum | 6–30 | 0 | | 100 |
| Rice | 2–10 | 17 | | 83 |
| Sweet potato | 10–25 | 20 | | 80 |
| Sago palm | 15–65 | 27 | 740 | 83 |
| Banana | 5–90 | 20 | | 80 |

combination with starch. However, they tend to be brittle and somewhat water sensitive, although the brittleness can be overcome by introducing some form of plasticiser and the water sensitivity can be controlled by cross-linking. Alginates are the main skeletal polysaccharides of the brown algal seaweeds. They have a linear block co-polymer structure of α-L-guluronic and β-1,4-D-mannuronic acid units (Figure 8.5) and they exist naturally as their sodium, magnesium and calcium salts, the latter assisting in cross-linking the gel stucture. They are initially biosynthesised as chains of β-1,4-D-mannuronic acid and this is followed by epimerisation of blocks of residues at C-5 to α-L-guluronic acid.

Of the synthetic polymers used for surface sizing, polyvinyl alcohol is the most important and tends to be used for films which require high tensile strength, high transparency and good oil resistance. It is made by the hydrolysis of polyvinyl acetate. Some residual acetate groups are usually present and these influence the level of water solubility (the lower acetyl content somewhat surprisingly being less soluble); the extent of hydrolysis is generally between 88% and 99%. They are available in a variety of molecular weights and acetyl content which give a range of rheological and film-forming properties.

Sometimes internal sizing agents such as alkyl ketene dimers and

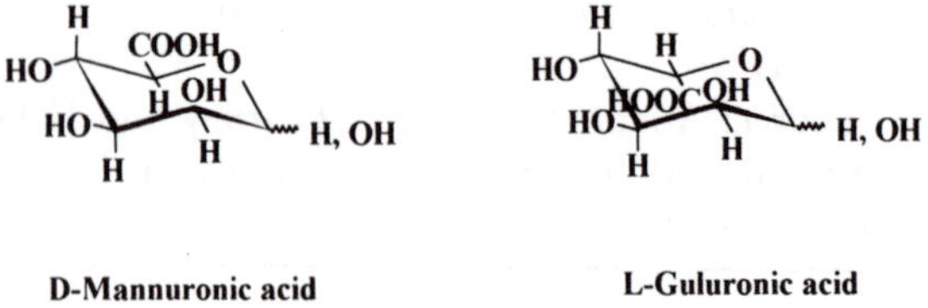

**D-Mannuronic acid**                              **L-Guluronic acid**

**Figure 8.5**   D-*Mannuronic and* L-*guluronic repeat units of alginic acid*

alkenyl succinic anhydrides (as discussed in Chapter 7) are used in conjunction with the surface sizing polymer to assist in improving the coating hold-out. This is not normally necessary when these internal sizes have been used in the formation of the base sheet.

## PIGMENT COATING

Although pigment coating provides a surface which completely covers the fibres of the base paper, it does not hide defects in a poor base sheet. The base paper therefore has an important role in determining the ultimate quality of the coating. It should have good formation and smoothness, otherwise a higher, and therefore more costly, coat weight will be necessary to cover the surface. It should also have carefully controlled surface porosity and strength to allow it to run effectively on high speed coating machines.

### Coating Methodology

Coating is performed by applying a relatively high concentration of a dispersion of a pigment (about 60% w/w) to the surface of a base paper. The coating dispersion is applied by metering on to the paper surface to give the correct coating weight, normally expressed as a weight per unit sheet area. Medium- to light-weight coats are around $15 \, \mathrm{g \, m^{-2}}$, and above this the coat weight is considered to be heavy.

A large variety of methods exist by which pigment coating mixtures can be applied to paper, and it is beyond the scope of this chapter to discuss these in detail. However, some general description of the principles will be helpful. There are two basic methods of pigment coating: blade coating and air-knife coating. In the blade coating procedure, a flexible metal blade, which is virtually in contact with the paper surface, is used to apply the coating mix. The angle of the blade and the application pressure are used as the primary control variables. In the air-knife coating system an excess of the coating dispersion is applied to the paper surface and the excess is removed by applying a well controlled stream of air (the air-knife) from a slot located near the paper surface. Air-knife coaters usually require a coating mix with a lower viscosity and therefore a lower solids content than blade coaters. The coated sheet is then dried using either infrared or thermal procedures. A diagrammatic representation of the two methods is shown in Figure 8.6.

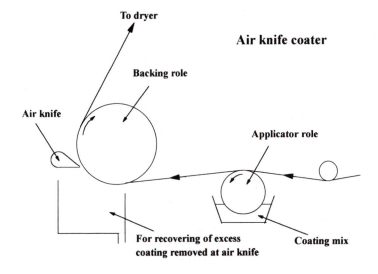

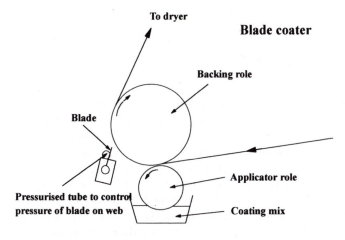

**Figure 8.6**  *The principles of the air-knife and blade coater.*
(Source: kindly drawn by Dr. D.J. Priest, Dept. Paper Science, UMIST).

Modern coating is performed at very high speeds, and the fastest coating operations at the moment are carried out at speeds approaching 1600 metres per minute. The rheology and flow behaviour of the dispersions at these speeds is therefore of paramount importance in the coating process.

## Coating Chemistry

In order to prepare a coating mixture (or coating colour as it is more usually known), a binder is added to a pigment suspension. The binder may be in the form of an emulsion or a solution, and its function is to provide bonding between the pigment particles and also between the pigment and the base paper. The binder also alters the rheological properties of the coating mix. It is usually necessary to achieve as high a solids content as possible for a given viscosity of the coating mixture, and in order to do this, the suspended particles must have a high degree of colloidal stability, which in turn is achieved by controlling the particle size distribution and by adding deflocculants.

The most common pigment is kaolin clay, which is often calcined or chemically aggregated to increase its coating bulk and porosity and to improve its general ability to give good base sheet coverage. The other important pigment is calcium carbonate which may be either natural ground carbonate or, increasingly commonly, precipitated calcium carbonate. Precipitated calcium carbonate is produced by controlled precipitation with carbon dioxide from calcium hydroxide solution, which allows particles of differing crystalline and morphological form to be produced. Less commonly used are gypsum, amorphous silica and titanium dioxide for rather more specialised applications. The particle size of the pigment is usually around $2\,\mu$m (expressed as an equivalent spherical diameter—see Chapter 6).

The binders vary quite widely—the most common being starch, soy protein and latexes in conjunction with other soluble polymers. Styrene–butadiene latexes have been the most popular but ethylene–vinyl acetate binders are also used. The method of polymer synthesis provides a way of modifying the properties of the latex. For example, adjustment of the ratio of styrene:butadiene in the co-polymer gives rise to different degrees of softness or hardness. This property has a profound influence on the quality of the coating. It is also possible to co-polymerise monomers so as to introduce, for example, carboxy groups on to the surface of the latex particle which in turn assist in colloidal stability. This avoids the use of excessive surfactant addition in the dispersions, the use of which can sometimes have adverse effects on coating properties. The strength of the adhesion required depends upon the kind of printing process which is to be used. In some forms of lithographic printing, inks are transferred to the surface of the paper under pressure from a rubber covered roll, and the coating layer must be sufficiently well-bonded to prevent disintegration of the surface.

The other important function of the binder is its effect upon the rheological properties of the coating mix. Starch, which is widely used, is unsuitable for use in its unmodified form because its solution viscosity is generally too high and also because of the problem of retrogradation. It is usually modified by reducing its molecular weight by either oxidative or hydrolytic (sometimes enzymatic) procedures.

In addition to these two primary components of the coating dispersion, a number of other chemicals are also often used, the most important of which are dispersants, flow modifiers, water retention agents and insolubilisers. Dispersants are usually polymeric and are used to inhibit the tendency of the dispersion to flocculate. Flow modifiers are normally water-soluble polymers (often cellulose derivatives, in particular carboxymethyl cellulose) which are used to modify the viscosity of the dispersion. The viscosity modification may arise not only from the viscosity of the polymer solution but also from the adsorption of the CMC molecules onto pigment surfaces. The function of insolubilisers is to avoid loss of strength of the coating layer during a printing process in which wetting of the surface is involved, and this is a particular problem with the very hydrophilic binders such as starch. The insolubilisers are cross-linking agents often based on urea–formaldehyde and melamine–formaldehyde resins, which are added to the coating mix as low molecular weight solutions and which assist in giving a greater level of water resistance to the dried coating.

## The Rheology of the Coating Mix

The flow behaviour of aqueous coating dispersions, because of their high pigment and binder content, is often complex. They have viscosities which are not independent of the shear rate and are therefore non-Newtonian. Shear thickening (when the viscosity of the dispersion increases with shear rate) and shear thinning or pseudoplastic behaviour (when the viscosity decreases with shear rate), may both be encountered. Shear thickening is a particularly common problem in coating and is usually encountered at high solids concentrations, particularly for clay suspensions where the disc-shaped particles are not completely free to rotate. The effect can sometimes be initiated by water being lost into the base paper leading to a localised solidification of the suspension and giving rise to uncoated areas of varying size on the paper.

*Chapter 9*

# Recycling of Cellulose

## INTRODUCTION

Recycled paper is a major source of fibre for the pulp and paper industry, and now provides around a third of the world's fibre for paper and board production (Table 1.2, Chapter 1). This figure is still rising, albeit only slowly. Recycling of paper is generally considered to be environmentally beneficial, but the environmental arguments are not as straightforward as might be imagined and these are discussed more fully in Chapter 10. In Europe, where there is a wood deficiency, recycled fibre is an important cheap and readily available resource, and the driving force for increased levels of recycling has therefore been mostly economic, although consumer pressure to recycle is now considerable. In North America and Scandinavia, where wood is plentiful, recycling has not been such an economic necessity. Nevertheless, consumer pressure has been responsible for a significant increase in recycling over the last few years but still at a much lower level than in Europe.

This chapter is concerned not with the environmental issues, which are discussed more fully in the following chapter, but with the technical problems associated with the recycling of cellulosic fibres. The quality of paper made from recycled fibre is generally less good than that of the unrecycled (virgin) material, but the behaviour of different types of pulp can be very variable in this respect, and a full understanding for the causes of these effects is still lacking. In this chapter some of the more important technical aspects of the chemistry of fibre recycling are discussed.

## GRADES OF WASTE PAPER

The grading and collection of waste paper are not well-standardised procedures. Some countries concentrate on the recovery of papers in particular grades whereas others concentrate on recovery in terms of volume. However, techniques for collection are continually evolving.

The most common grade of waste paper is what is often called 'mixed waste' and this is made up of various qualities of paper. It usually contains a high proportion of newsprint and packaging grades of board and it is used primarily for low-cost products such as board. The second important grade is 'packaging waste' which is usually mostly corrugated packaging board consisting of double-lined kraft pulp as an outer surface with a fluted medium centre. This grade is used predominantly in the production of liner-board, new corrugated board and in dry-wall board. In addition to these grades there is also 'unprinted writing' which is of high quality and normally contains only white paper with a reasonable quality of brightness and having no printing. Finally there are printed papers which are recovered from newsprint, magazines or office waste and these must usually be deinked prior to use.

## Collection and Recovery

Packaging waste is normally recovered from one of two sources. The first is the waste which is not used by the converter (*i.e.* the maker of the package) and the second is from the packages which are discarded at the point of sale. The former is relatively free of contaminates whereas the latter may contain adhesives, plastics, staples and other 'contraries'. Many countries, including the UK, have developed very efficient collection and recovery systems for packaging material from wholesale and retail outlets. For example in 1991, the Japanese were recovering almost 70% of corrugated containers from this source.

Recovery of newsprint and magazines follows a similar pattern. The waste from printers for example from the edge trim or from the end or reels, *etc.*, is normally unprinted and therefore commands a high price. Unsold or undistributed newspapers are also returned to the distributers and are used for recycling, but these will usually be deinked. The recovery of used newspapers and magazines is a more difficult problem than waste packaging material because they are so much more widely distributed. Many households and businesses now separate paper from their waste and take it to central collection points. However there is still a substantial amount of paper discarded with domestic refuse, and about 20–25% of domestic refuse is still made up of paper—a substantial proportion of which is newspapers and magazines. However, as awareness of the import-ance of recycling has continued to develop, the proportion of newspapers and magazines discarded as refuse has been decreasing. In Japan in excess of 98% of newsprint is now recovered, a level of recovery which is achieved by domestic collection of all grades which are pre-sorted by residents.

High quality printing and writing grades are the best source of high quality fibre. These are usually free of mechanical pulps and are often known, somewhat confusingly, as 'wood free' papers. Printers and converters usually waste around 5–15% of their raw material as unprinted off-cuts through cutting and other converting processes, and this is available for recycling. It is also becoming increasingly common for offices and businesses to recycle paper, and also for these papers to be sorted in terms of quality (for example: coloured and non-coloured; printed and unprinted). Many pilot studies have been carried out into the availability of paper from such sources and these show that between 1 and 5 kilogrammes of paper per week per office occupant might be expected to be available. Ten office workers would therefore produce, on average, a tonne of paper per year and the potential for recycling from this source is clearly very considerable, and in many countries is under-exploited.

## CHANGES IN PAPER AND FIBRES DURING RECYCLING

### Changes in Mechanical Strength and Other Properties

For pulp which has been chemically delignified and which has then been refined, recycling causes a major reduction in tensile, bursting and folding strength, and a lesser reduction in apparent density and stretch. In addition, there is usually an increase in tearing resistance, stiffness and air permeability. Light scattering properties such as opacity also increase. The biggest effect occurs on the first occasion that the delignified pulp is dried and recycled. Strictly, therefore, dried chemical wood pulps should be considered to be once-recycled fibres. Some of these changes are shown in Figure 9.1.

Mechanical pulps (*i.e.* those in which the lignin is still largely present) in contrast to chemical pulps do not show a very significant loss in strength on recycling. However, it should be appreciated that, in the first place, these pulps are substantially weaker than chemical pulps. A comparison of changes in chemical and mechanical pulps during recycling is shown in Figure 9.2.

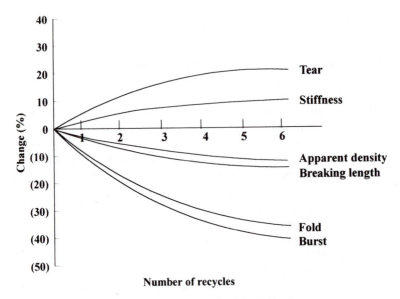

**Figure 9.1** *Effect of repeated recycling on refined chemical pulps.*
(Source: R.C. McKee, *Paper Trade J.*, 1971, **155**(21), 34).

## Fibre Characteristics

Figure 9.2 shows that, during recycling of chemically delignified (Kraft) pulp fibres, irreversible pore closure within the cell wall takes place which leads to a reduction in their cell wall water content as measured by the fibre saturation point (see Chapter 5). The net effect of this is a loss in fibre flexibility which, in turn, leads to less effective inter-fibre bonding.

There is much supporting evidence for the theory that loss of strength during recycling is due to the fibres becoming less flexible in their wet state and also to their surface properties changing. Both of these would be influential in bonding the cellulosic surfaces together. It is also known that fibres become more brittle on recycling, and this, combined with their poorer flexibility in the wet state, makes them less able to achieve the conformation required for optimal bonding.

The fibres themselves do not seem to suffer any particularly significant reduction in strength during recycling, and to some extent this explains the somewhat curious fact that the resistance of paper to tearing improves during recycling (Figure 9.1). It is well known that, as the bonding of the sheet of paper becomes better, its

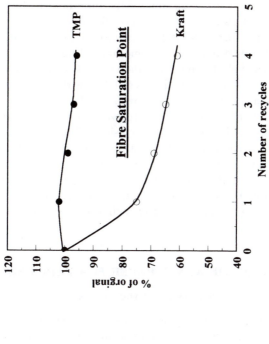

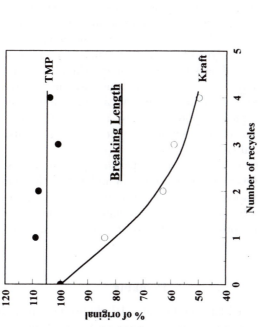

**Figure 9.2** *Comparison of the effect of repeated recycling of Kraft and thermomechanical pulp (TMP) fibres on tensile strength and fibre swelling.* (Source: Adapted from R.C. Howard and W. Bichard, *J. Pulp and Paper Sci.*, 1993, **19**(2), J57 and 1992, **18**(4), J151).

resistance to tearing falls. The improvement in tear resistance may therefore arise from the weakening of the fibre bonds which in turn helps to absorb the stresses of tear more effectively.

The problem of irreversible pore closure in the cell wall is more serious in delignified pulps than in mechanical pulps because, in the former, the removal of lignin from the cell wall allows the cellulosic polysaccharide surfaces to bond irreversibly together. This phenomenon of 'hornification', as it is known, is also observed in other polysaccharides, most notably starch, which undergoes irreversible retrogradation on drying. Solute exclusion (fibre saturation point) methods have been used to demonstrate that the size of the large and intermediate pores in the fibre cell wall are reduced upon drying. The possibility of chemical cross-linking *via* hemiacetal formation between the cellulose and hemicellulosic chains has been proposed, but this is speculative and such a mechanism has yet to be proved.

## PREPARATION OF WASTE PAPER FOR PAPER MAKING

### Dispersion of the Fibres

The first step in recycling waste paper involves the mechanical dispersion of the paper into its component fibres in water at about 5–10% consistency. This is followed by a series of cleaning operations which are designed primarily to remove non-fibrous contaminants such as metal, elastic and adhesives. These cleaning steps may be either physical or chemical and usually involve screens and centrifugal cleaners. Adhesives which are present in bindings and which are frequently polymeric are amongst the most difficult contaminants to remove. Surfactants and chelating agents, however, have a beneficial effect.

It is common during the dispersion process to use sodium hydroxide as this assists in ink removal and also in the more efficient dispersion of papers which have been chemically sized (see Chapter 7). Sodium hydroxide also has a swelling effect on cellulosic fibres and it is most probable that this too has a beneficial effect on strength.

### Deinking

Many papers used for recycling are already printed upon, and one of the most important chemical processes involved in waste paper preparation for high quality grades is the removal of ink. About 15% of all recycled waste paper needs to be treated to remove ink, and the deinking process accounts for most of the chemical use in recycling. World-wide, the deinking of waste paper has grown dramatically in the past fifteen years from around 3 million to 15 million tonnes per annum.

The first stage in deinking is normally the adjustment of the fibre suspension to a high pH. This causes some swelling of the fibres which, together with mechanical action, assists in loosening the print particles. The alkali assists in saponifying the varnish or the vehicle of the printing ink thus allowing the release of the pigment contained in the ink. As some print consists of pigment particles held together by cross-links, the sodium hydroxide may also cleave these cross-links and help to break up the print into smaller particles. A detergent is also usually added in order to assist in the dispersion of the pigment and to prevent its agglomeration after release from the paper. Once the ink pigment has been released, it is necessary to prevent it from becoming re-deposited onto the surface of the fibres which have a strong affinity for pigments. Low levels of bentonite (about 0.75%) are typically used to absorb the ink pigments. The alkali is also helpful in breaking down the sizing chemicals in the paper which are often bonded *via* alkali-sensitive ester links. In the case of rosin sizing the alkali assists in desizing the sheet by saponification of the rosin acids. The alkali normally used is either sodium carbonate or sodium hydroxide, the former being milder and giving rise to less oxidation of the fibres and less fibre loss. However, a faster rate of dispersion can be achieved with sodium hydroxide solution.

The ink particles are then removed by either washing or flotation. The simplest form is wash deinking which consists of sequentially dewatering several times. The washing process will normally improve the brightness of the stock if the print particles which are present are in the size range 1–10 $\mu$m in diameter.

For more efficient deinking, flotation must be used. This is considerably more complex, and the mechanism by which it works is still somewhat speculative. It is necessary to make the detached ink particle hydrophobic so that it can adhere to an air bubble and be removed by flotation. For this to happen, it appears to be necessary for the particles to be in the size range 10–150 $\mu$m. The processs consists of a cell or tank with a high speed agitator to induce a partial vacuum which causes air to enter the system in the form of small air bubbles, these carry the ink particles to the surface.

Typically around 3–5% of saponified fatty acids would be added to the flotation cell. The air bubbles attract ink particles and, as the ink-rich air bubbles reach the surface, they are removed by means of a slowly rotating paddle. The froth is withdrawn and further concentrated prior to disposal.

The saponified fatty acids which are used are most often palmitic, stearic or oleic acid but the way in which they confer a hydrophobic nature to the surface of the ink particle is not well understood. If enough calcium ions are present (and these sometimes need to be added to the system) insoluble calcium salts of the fatty acids are probably produced and these may coat the surface of the print particle making it hydrophobic. The ink particle then adheres to an air bubble and can be floated out of the stock. The saponified fatty acids are often called 'collectors'—a term which comes from mineral flotation.

The brightness of the deinked fibres may be further improved by bleaching. This may be done by use of hydrogen peroxide, sodium hypochlorite, or by some form of chlorination. In the case of hydrogen peroxide, silicates and chelating agents are usually necessary in order to inactivate metal ions which are almost always present in waste and which would otherwise be able to decompose the peroxide catalytically. Deinked stock is, nevertheless, normally less white than virgin fibre and is often used in mixture with virgin pulp. Recycled pulp requires no refining and is therefore added after the refining of virgin pulp with which it is to be mixed.

*Chapter 10*

# Paper Making and the Environment

## INTRODUCTION

As a major user of a photosynthetically renewable and recyclable raw material and of water which must finally be discharged to the environment, it is hardly surprising that environmental matters have a high profile in the paper industry. However, these issues are complex and the debate all too often suffers from oversimplification. This chapter attempts to identify and clarify some of these issues, particularly where the chemistry of the process is involved.

## THE FIBRE RESOURCE

### Natural and Plantation Forests

One of the most debated environmental issues of the past fifteen to twenty years has been the exploitation of the forests for wood for paper making. Approximately 30% of the earth's land surface is forested, and around half of this is harvested commercially for industrial purposes (Chapter 1). Over 80% of this wood for industrial use comes from the forests of North America, Europe and what was formerly the Soviet Union. Wood has been the primary fibre source for pulp and paper production world-wide for many years, and it is necessary to take a global view of its consumption. Wood consumption world-wide has more than doubled since 1950 from 1.5 billion to 3.5 billion $m^3$ (United Nations Food and Agricultural Organisation). Approximately half of this is used for fuelwood and half for industrial use. The principal driving force for this increase in consumption has been the increase in world population which shows a close correlation with wood consumption (Figure 10.1).

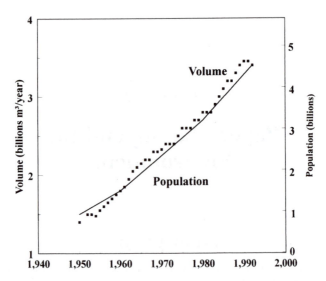

**Figure 10.1**   *Wood consumption and world population 1950–1995.*
(Source: Redrawn with permission from 'Fibre Supplies of the
Paper Industry', W. Sutton, Seminar notes, UMIST, 1994, 4).

The annual per capita wood consumption is now around 0.67 m³
per person per year and this figure has only increased slightly since
1950. Based upon predictions of future world population growth, the
'average annual increase' in wood demand is therefore likely to be
around 86 million m³. This 'annual' global increase is, astonishingly,
slightly more than the current total annual harvest of British
Columbia. This total, of course, includes fuelwood and, if only
industrial wood is considered, then it would be necessary to find the
equivalent of the wood harvested in British Columbia every two
years. This is clearly an alarming environmental problem.

It is a common misconception that most of the wood extracted for
industrial use is used for making paper, and it is therefore equally
incorrect to conclude that there would be very much impact upon
the world's forests if more paper were to be recycled. Over 80% of
the wood harvested industrially in the world is used for the
manufacture of sawn timber, plywood and other solid wood products. Because not all of the felled tree is suitable for this purpose,
paper is generally made either from logs which are unsuitable for
sawing or peeling, or from residues arising from these processes.
Very few forests are harvested solely for paper making, and recycling
will therefore have, at best, a minimal impact in reducing wood
harvesting levels. The paper industry should therefore be viewed

more accurately as a user of forestry waste rather than as a primary consumer of wood. Some of the increased need for wood will be satisfied from traditional natural forests but there is growing environmental pressure to find alternatives and, if the demand is to be satisfied, this must come from plantation supply. At the moment, globally, there are about 100 million hectares of plantations, the vast majority of which (over 85%) consist of slow growing species in the Northern Hemisphere. Many of these plantations are under 30–35 years old. Plantations in the Southern Hemisphere would be potentially more productive. For example, one hectare of managed *Radiata pine* grown in New Zealand produces as much as 40 hectares of the Amazon rain forests. However, current estimates are that not more than 20–25 million m$^3$ of sawlogs can be expected above 1990 levels from plantation sources.

There are not strong reasons therefore for the paper industry to move away from wood as a raw material and, although alternative sources of fibre will probably enjoy a greater share of the pulp market, these will probably continue to be of relatively minor importance.

## Recycling

Recycling clearly has an important bearing upon fibre supply. There are two important environmental aspects to waste paper recycling. The first of these is what is known as the utilisation rate of waste paper and is defined as the waste paper consumption as a proportion of total paper and board production. The second useful definition is the recovery rate, which is simply the amount of waste paper recovered as a percentage of total paper and board consumption. An example of the use of such figures is shown in Table 10.1.

When these figures are compared to those of other major industrial countries (Figure 10.2), it is clear that the United Kingdom has a relatively good utilisation rate (57%) but its recovery rate is close to the bottom of the international league table (30%).

The chemistry and technology of recycling are discussed in Chapter 9 and it is intended in this chapter only to discuss the environmental effects. These are generally considered by the public consumers of paper and board to be good, but there are environmental disadvantages which must be offset against the obvious environmental advantages of reduced wood consumption and reduced need for chemical pulping processes. The main disadvantages are that the recycling of printed papers often requires the use of a

**Table 10.1** *Waste paper recovery and use in the UK (1985–1989).*
(Source: N. Kirkpatrick, 'Environmental Issues in the Pulp and Paper Industries, Literature Review', Pira Reviews of Pulp and Paper Technology, Pira International, 1991, p. 20).

|  | *1985* | *1986* | *1987* | *1988* | *1989* |
|---|---|---|---|---|---|
| Paper and board production (millions of tonnes) | 3.77 | 3.94 | 4.18 | 4.29 | 4.48 |
| Waste paper used (millions of tonnes) | 2.07 | 2.45 | 2.31 | 2.42 | 2.58 |
| Utilisation rate (%) | 54.9 | 54.5 | 55.2 | 56.3 | 57.6 |
| Total paper and board consumption (millions of tonnes) | 7.80 | 8.07 | 8.73 | 9.29 | 9.59 |
| Waste paper recovery (millions of tonnes) | 2.17 | 2.36 | 2.60 | 2.78 | 2.98 |
| Recovery (%) | 27.8 | 29.3 | 29.8 | 29.9 | 31.0 |

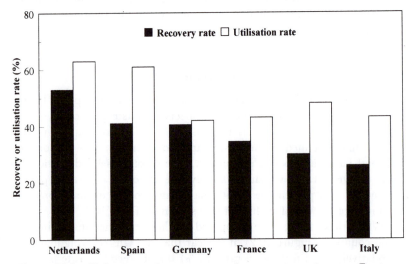

**Figure 10.2** *Utilisation and recovery rates for waste paper in some European industrialised countries.*
(Source: N. Kirkpatrick, 'Environmental Issues in the Pulp and Paper Industries, Literature Review', Pira Reviews of Pulp and Paper Technology, Pira International, 1991, p. 21).

deinking process which produces an ink-rich waste slurry which must then be disposed of, and the dissolution of organic nutrient, particularly from starch, which imposes a burden on the effluent treatment system.

# WASTE DISPOSAL

## Water Use

The production of paper requires very large quantities of water. In the United Kingdom, the average water consumption is around 20–30 m³ per tonne of paper and, as the industry produces around 4 million tonnes of paper per year, this represents an annual water consumption of about 80–120 million m³ a year. Water is abstracted from inland streams, from bore holes or, in some exceptional cases, mains water may be used. The abstraction is carefully controlled legally (in the UK by the National Rivers Authority) and, after use in manufacturing, it is usually returned to the aquatic environment by way of discharge to an inland stream, an estuary or to the sea. The industry has reduced its water consumption substantially over the past 25 years and is continuing to do so. It is possible in some circumstances, by appropriate internal recycling of water, to operate a paper production unit with zero effluent discharge. However, some processes, particularly those involving the production of coloured grades, will always be difficult to run at very low levels of water consumption.

## Waste Water Discharges

Discharges from the paper industry to the aquatic environment may be either nutritionally polluting or toxically polluting or both, and the nature of the pollution is very dependent on whether or not the process involves either delignification or bleaching. Pulping (delignification) and bleaching are performed almost exclusively outside the UK, and produce highly polluting effluents. In the Kraft (Sulfate) process, the black liquor from delignification, in a properly operating system, should not be discharged but should be completely recovered and reused by concentration, combustion and dissolution of the inorganic chemicals. However, aquatic pollution can sometimes occur as a result of spillages. The sulfite process, on the other hand, does not normally have a chemical recovery system associated with it and the soluble lignosulfonates produced from the delignification process must be discharged. The disposal of bleachery waste is a more difficult problem. Effluents from chlorine-based bleaching processes tend to be dilute but contain hazardous organochlorine compounds and it is therefore necessary to treat them in some way to detoxify them prior to their discharge to the aquatic environment.

## Nutrient Pollution

The production of paper from chemical pulp, or from recycled fibre where no delignification is involved (as in most UK operations), generates an effluent which generally has little toxicity but which does contain biodegradable nutrients. These biodegradable organic molecules are mostly derived from carbohydrate and are utilised for growth and respiration by the micro-fauna and -flora in the stream thus leading to a reduction in dissolved oxygen. If this happens to a serious extent the ecological balance of the stream may be affected, particularly oxygen-sensitive species and, in the worst cases, total deoxygenation and the onset of anoxic (oxygen free) conditions may ensue. The existence of anoxic conditions may then give rise to toxicity *via* the formation of species such as sulfide ion, which can be produced by reduction of sulfate ion by sulfate-reducing bacteria. The dangers of this type of problem are quite significant in paper-making effluents as they often contain high concentrations of sulfate ion.

The main nutrients which are present in a paper mill effluent are carbohydrates from wood pulp or from waste paper, and also products which arise from their degradation. Compositions may be very variable and depend to a large extent upon the type and amount of waste paper, if any, which is being used. Recycled coated and surface-sized papers will introduce significant amounts of soluble polymers, in particular starch, which were used in the original surface treatment.

The effect which nutrient discharges have on the dissolved oxygen of a stream is best demonstrated by reference to a simple organic molecule. If the total oxidation of glucose to carbon dioxide and water is considered (equation 10.1), one part by weight of glucose would require 1.06 parts of oxygen for complete oxidation.

$$C_6H_{12}O_6 + 6O_2 \rightarrow 6CO_2 + 6H_2O \qquad (10.1)$$

As the solubility of oxygen in a stream which has been saturated with air is around 10 milligrams per litre at about 12 °C, 1 gram of glucose, if completely oxidised in the aquatic environment, would require theoretically 106 litres of water to provide the oxygen. This is what is known as the theoretical oxygen demand (TOD) of glucose and gives a rough measure of the dilution necessary in a stream. It can be determined experimentally by subjecting the organic molecule to complete oxidation by refluxing with acidified dichromate and determining the equivalent oxygen consumed by volumetric analysis. This measured value is known as the chemical oxygen demand (COD) and usually agrees well with the theoretical value. Because most effluents are mixtures of unknown composition, the theoretical calculation is not possible and the chemical oxygen demand is usually determined experimentally.

However, complete oxidation to carbon dioxide and water is not achieved under natural stream conditions and a more useful measurement is the deoxygenating effect of a discharge under simulated stream conditions—this is known as the biochemical oxygen demand (BOD). It is determined experimentally by diluting the effluent in simulated stream water which, like normal stream water, contains a community of microorganisms (a microbiological seed) and various essential inorganic nutrients and trace elements. The drop in dissolved oxygen (the oxygen demand) which occurs over a period of five days is then measured. The experiment must be performed in a sealed stoppered bottle with no trapped air and in the absence of light to avoid the formation of oxygen *via* photosynthesis. For 1 gram of glucose the BOD is around 0.6 grams of oxygen (as compared to the theoretical maximum oxygen requirement of 1.06). A comparison of the theoretical, chemical and biochemical demands of some common chemical substances is shown in Table 10.2.

The experimental determination of COD for most organic compounds agrees well with the theoretically predicted value (TOD) but the BOD is almost always lower. This is to be expected, because complete oxidation does not take place in the aquatic environment. The COD:BOD ratio can therefore be regarded as a measure of aquatic biodegradability. The closer this value is to 1 the greater the aquatic biodegradability of the compound. In industrial nutrient discharges it is usually the case that the effluent contains a mixture of nutrients of uncertain composition. It is therefore not possible to

**Table 10.2** *A comparison between the TOD, COD and BOD of some common nutrient pollutants.*

| | | TOD (g/g) | COD (g/g) | BOD (g/g) | COD/BOD |
|---|---|---|---|---|---|
| Glucose | $C_6H_{12}O_6$ | 1.06 | 1.10 | 0.60 | 1.83 |
| Acetic acid | $CH_3COOH$ | 1.06 | 1.05 | 0.50 | 2.10 |
| Phenol | $C_6H_5OH$ | 2.38 | 2.24 | 1.80 | 1.24 |
| Formic acid | $HCOOH$ | 0.34 | 0.34 | 0.20 | 1.70 |

determine a theoretical oxygen demand, and so the COD is measured experimentally. The determination of chemical or biochemical oxygen demand, in conjunction with a knowledge of the solubility of dissolved oxygen in water, allows an estimate to be made of the effect of the discharge upon the receiving stream. A simulation of the effect on a stream of the discharge of a nutritionally polluting effluent is shown in Figure 10.3.

Although the BOD of the stream increases almost immediately after the discharge, the dissolved oxygen takes some time to reach a minimum value, and the maximum impact is often experienced many miles downstream of the discharge. The stream then gradually recovers as a result of natural re-aeration via turbulence and photosynthesis.

Some forms of nutrient pollution may be found under conditions where deoxygenation is not a serious problem, and this has often been observed at the site of paper mill discharges. These forms of pollution manifest themselves as a voluminous growth of filamentous bacteria and they sometimes occur as effluent quality is improving.

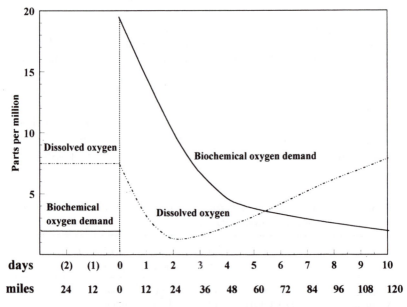

**Figure 10.3**   *Model of the effect of a nutrient (domestic sewage) discharge to a stream.*
(Source: 'Industrial Pollution'; ed. N.I. Sax, Van Nostrand Reinbold, New York, 1974, p. 206).

These effects can be observed in some cases at extremely low nutrient doses. These growths are often dominated by the sheathed filamentous bacterium, *Sphaerotilus natans*, which has the ability to attach itself to submerged surfaces and abstract nutrients, in particular low molecular weight carbohydrates, at very low concentration — much lower concentrations indeed than those which would give rise to any significant increase in oxygen demand of the stream. *S. natans* is a member of the group of self-purifying bacteria which are always present in streams and whose role, under normal conditions, is to remove trace organic nutrients.

## Toxic Pollution

The major cause of toxic pollution in the past by the paper industry has been from discharges of bleachery waste from chlorine-based bleaching processes. These processes produce a wide variety of chlorinated-organic molecules of varying toxicity. The wood resin acids, which are removed from wood during pulping, have also been found to be toxic to fish, and although these are largely recovered as by-products in chemical pulping, some may still be found in pulp mill effluents.

Until recently, chlorine has been the dominant bleaching process, but its use is diminishing rapidly. Effluents from chlorine-based bleaching processes are relatively dilute and chemical recovery and reuse is impractical. Most of the organochlorine compounds formed during chlorine bleaching arise from the chlorination of residual aromatic residues of lignin in the pulp. It has been estimated that, for bleached pulps, up to 90 kg of organochlorine compounds may be produced per tonne of pulp and that between 75 and 90% are released during the chlorination and subsequent alkaline extraction stages. In addition, organochlorine compounds may have been added directly to the wood or the wood chips prior to pulping to control microbiological growths. Some chlorine-containing organic compounds may also be added during the wet formation part of the paper-making process itself (Chapter 7). Many of the organochlorine compounds found in bleachery waste exhibit both toxicity and mutagencity to aquatic organisms (Table 10.3).

Chromatographic separation and mass spectrometry are necessary to identify and measure quantitatively individual organic halogen components, but these are not practical for routine assay, and more general measurements are usually used. The first is the measurement of total organohalogen (TOX), and the second is of adsorbable organohalogen (AOX). These methods are based upon the principle

**Table 10.3**   *Occurrence and toxicity of chlorinated phenolics from the bleaching of hardwood and softwood pulps.*

| Structure | Name | 96 h $LC_{50}$ (mg/l) | Total chlorinated phenolics (g/over-dried tonne) | |
|---|---|---|---|---|
| | | | *Hardwoods* | *Softwoods* |
| | Dichlorocatechol | 2.7[1] | | |
| | Trichlorocatechol | — | 15–43[3] | 72–260[4] |
| | Tetrachlorocatechol | 0.8[1] | | |
| | Dichlorophenol | 1.7[1] | | |
| | Trichlorophenol | — | | |
| | Tetrachlorophenol | 0.48[1] | | |
| | Dichloroguaiacol | — | | |
| | Trichloroguaiacol | 0.75[2] | | |
| | Tetrachloroguaiacol | 0.32[2] | | |

Sources: 1 BC Research, CPAR Report 245-1, 'Identification of the Toxic Constituents in Kraft Mill Bleach Plant Effluents'. May 1974, p. 22. 2 BC Research, CPAR Report 245-2, 'Identification of the Toxic Constituents in Kraft Mill Bleach Plant Effluents'. May 1975, p. 31. 3 R.H. Voss, J.T. Wearing and A. Wong. 'Effect of Hardwood Chlorination Conditions on the Formation of Toxic Chlorinated Compounds', *Tappi*, 1981, **64**, 3, pp. 167–170. 4 R.H. Voss, J.T. Wearing and A. Wong, 'Effect of Softwood Chlorination Conditions on the Formation of Toxic Chlorinated Compounds', 1979 CPPA/TS Environmental Conference, Victoria BC, Canada, November 1979.

of adsorption of organics from water on to granular activated carbon packed in microcolumns. Inorganic halides are then removed from the column by washing with nitrate solution. The remaining sorbed organics are then combusted together with the granular activated carbon and the halides determined by microcoulometric titration with silver ion. TOX measures the organohalogen content of an unfiltered sample whereas the AOX measurement is made on a filtered sample. Organic halogens do not occur in natural waters at concentrations above approximately 5 $\mu g\,l^{-1}$ and a TOX above this level is usually indicative of contamination by synthetic organohalogen compounds. Fluorinated organics do not produce species which are not titratable against silver ion and are not therefore detected by the test.

Whilst these measurements give a useful measure of the level of chlorinated organics, they are not particularly helpful in assessing toxicity which may vary widely between individual components. At the moment AOX and TOX limits are used for the control of discharge of chlorinated organics in a number of countries, but it is likely that future legislation may require the measurement of specific toxic chlorinated organics such as tri-, tetra-, and poly-chlorinated phenolics.

Dioxins are a particular group of chlorinated organic molecules which have been associated with pulp and paper production and are a concern because of their extreme toxicity. There are two groups of molecular types which fall into the general category referred to as dioxins. These are the polychlorinated dibenzodioxins (PCDDs) and the polychlorinated dibenzofurans (PCDFs). The structures of these molecules are shown in Figure 10.4.

There are 210 different isomeric possibilities, 75 of which are PCDDs and 135 are PCDFs. The toxicity of these isomers varies greatly, and only 15 exhibit extreme toxicity, the most toxic of which is 2,3,7,8-tetrachlorodibenzodioxin (2,3,7,8-TCDD). The toxicity of the other isomers is therefore expressed as a toxicity equivalent of 2,3,7,8-TCDD. The PCDDs and PCDFs are poorly water soluble but are fat soluble and are therefore able to accumulate in tissue fat, thus allowing them to bio-accumulate in living organisms. The origin of dioxins in the pulp and paper industry is not entirely clear. They may be produced from the chlorination of dibenzodioxin which may be present in recycled oils used to make defoamers, but they may also arise from wood chips which have been treated with polychlorophenol to prevent sap stain formation. It is also possible that they are derived from lignin by chlorination. Dioxins are also known to be formed naturally by combustion of material such as wood, and forest fires have been particularly identified as a likely major cause of dioxin emissions.

Dioxins have also been detected in paper products, albeit at extremely low levels. The mean values in one study for toilet tissue,

**Polychloro dibenzyl dioxin (PCDD)**        **Polychloro dibenzyl furan (PCDF)**

**Figure 10.4** *The structures of polychlorinated dibenzodioxins and polychlorinated dibenzofurans*

tea bags, sanitary towels, coffee filters and disposable nappies were 1.4 parts per trillion of 2,3,7,8-TCDD and 5.8 parts per trillion of 2,3,7,8-TCDF (tetrachlorodibenzofuran). These concentrations are around $10^6$ times lower than those which would represent a health risk.

The resin acids present in pulps, particularly those from softwood, have also been found to be significantly toxic to aquatic organisms. The amount in wood varies greatly between species (Table 10.4). Between 0.3 and 3.6 kg/tonne is extracted during pulping.

The distribution and toxicity of individual components are also variable (Tables 10.5 and 10.6).

**Table 10.4**   *Resin acid content of various woods.*

| Species | Total resin acids (kg/tonne of dry wood) | References |
|---|---|---|
| Lodgepole pine | 3.3–17 | 1, 2, 3, 4 |
| Loblolly pine | 3.9–12.3 | 1, 5 |
| Monterey pine | 6.9–15.2 | 2, 6, 7 |
| Southern pine | 3.3–6.5 | 8 |
| White spruce | 1.2–2 | 1, 2, 3 |
| Sitka spruce | 2.7 | 1 |
| Douglas fir | 2–4 | 1, 9 |
| Alpine fir | 0.3 | 10 |
| Western hemlock | 0.8 | 1 |
| Red cedar | 21 | 1 |

Sources: 1 D.H. Bennett, C.M. Falter and A.F. Campbell, 'Prediction of Effluent Characteristics, Use of Lime Treatments and Toxicity of the Proposed Ponderay Mill', Appendix in engineer's report on Effluent Characteristics for Washington State Department of Ecology, 1987. 2 J.M. Leach and A.N. Thakore, 'Toxic Constituents in Mechanical Pulping Effluents', *Tappi*, 1976, **59**, 2, pp. 129–132. 3 I.H. Rogers, A.G. Harris and L.R. Rozon, 'The effect of Outside Chip Storage on the Extractives of White Spruce and Lodgepole Pine', *Pulp and Paper Magazine of Canada*, 1971, **72**, 6, pp. 84–90. 4 Z. Wang, T. Chen, Y. Gao, C. Breuil and Y. Hiratsuka, 'Biological Degradation of Resin Acids in Wood Chips by Wood-inhabiting Fungi', *Appl. Environ. Microbiol.*, Jan. 1995, pp. 222–225. 5 D.F. Zinkel, 'Tall Oil Precursors of Loblolly Pine', *Tappi*, 1975, **58**, 2, pp. 118–121. 6 P.J. Nelson and R.W. Hemingway, 'Resin in Bisulphite Pulp from Pinus Radiata Wood and its Relationship to Pitch Troubles', *Tappi*, 1971, **54**, 6, pp. 968–971. 7 R.W. Hemingway, P.J. Nelson and W.E. Hillis, 'Rapid Oxidation of the Fats and Resins in Pinus Radiata Chips for Pitch Control', *Tappi*, 1971, **54**, 1, pp. 95–98. 8 D.F. Zinkel and D.O. Foster, 'Tall Oil Precursors in the Sapwood of Four Southern Pines', *Tappi*, 1980, **63**, 5, pp. 137–139. 9 D.O. Foster, D.F. Zinkel and A.H. Conner, 'Tall Oil Precursors of Douglas Fir', *Tappi*, 1980, **63**, 12, pp. 103–105. 10 M.F. McKenney, E.G. Adamek, G. Craig and J. Reinke, 'An Evaluation of Effluents Generated by a Thermomechanical Pulp Mill, presented at 1979 International Pulping Conference, Toronto, Canada, 1979.

**Table 10.5** *Distribution of resin acid components in various softwoods.*

| Resin acids (%) | Lodgepole pine[1] | Loblolly pine[2] | Monterey pine[3] | Douglas fir[4] |
|---|---|---|---|---|
| Pimaric acid | 6 | 8 | 6.8 | |
| Sandaracopimaric acid | 2 | 1.8 | 1.2 | 3.2–3.6 |
| Levopimaric acid | 7 | 59.5 | 49 | 2.3–6.2 |
| Palustric acid | 15 | | | 22.8–23.5 |
| Isopimaric acid | 15 | 1.6 | 2.5 | 27–27.8 |
| Abietic acid | 17 | 12.6 | 12.7 | 16.1–18.9 |
| Dehydroabietic acid | 25 | 8.4 | 8.1 | 7.8–10.1 |
| Neoabietic acid | 9 | 8.1 | 20 | 12.2–13.8 |

Sources: 1 D.H. Bennett, C.M. Falter and A.F. Campbell, 'Prediction of Effluent Characteristics, Use of Lime Treatments and Toxicity of the Proposed Ponderay Mill', Appendix in engineer's report on Effluent Characteristics for Washington State Department of Ecology, 1987. 2 D.F. Zinkel, 'Tall Oil Precursors of Loblolly Pine', *Tappi*, 1975, **58**, 2, pp. 118–121. 3 R.W. Hemingway, P.J. Nelson and W.E. Hillis, 'Rapid Oxidation of the Fats and Resins in Pinus Radiata Chips for Pitch Control', *Tappi*, 1971, **54**, 1, pp. 95–98. 4 D.O. Foster, D.F. Zinkel and A.H. Conner, 'Tall Oil Precursors of Douglas Fir', *Tappi*, 1980, **63**, 12, pp. 103–105.

**Table 10.6** *Toxicity of individual resin acids.*

| Resin acids | 96 h $LC_{50}$ (mg/l) |
|---|---|
| Dehydroabietic acid | 1.1 |
| Palustric acid | 0.5 |
| Abietic acid | 0.7 |
| Isopimaric acid | 0.4 |
| Pimaric acid | 0.8 |
| Neoabietic acid | 0.6 |
| Sandaracopimaric acid | 0.4 |
| Levopimaric acid | 0.7 |

Sources: 1 J.M. Leach and A.N. Thakore, 'Toxic Constituents in Mechanical Pulping Effluents', *Tappi*, 1976, **59**, 2, pp. 129–132. 2 L.T.K. Cheng, H.P. Meier and J.M. Leach, 'Can Pulp-mill Effluent Toxicity be Estimated from Chemical Analysis', *Tappi*, 1979, **62**, pp. 71–74. 3 J.M. Leach and L.T.K. Chung, 'Development of a Chemical Toxicity Assay for Pulp Mill Effluents', US EPA Publication, No. EPA-600/2-80-206, US Environmental Protection Agency, Cincinnati.

## Sludge Disposal

One of the major environmental problems facing the pulp and paper industry is the disposal of sludges. These arise from the settlement of paper mill effluents and they consist of particulate and short-fibred material which has passed through the machine-wire during the paper formation process. The effluents prior to discharge are usually treated to adjust their pH to between 6 and 8 and are then often treated with flocculants before passing into a settlement tank. The settlement process produces a clarified effluent and also a low consistency sludge (typically at about 5% solids). This sludge is usually thickened by rotary vacuum filtration to 20–30% or by centrifugation to a higher solids content (> 80%). The two major disposal methods for these sludges are either incineration, which is practised more widely in the United States and is best suited to sludges of high solids, or by land-fill, which tends to be used for wetter sludge cakes and is practised more widely in Europe. However, as regulatory control over land-fill becomes increasingly severe, costs are rising and it is doubtful whether land-fill will continue to be economic for many years to come. The sludges from these processes may vary considerably in their organic content, depending on the nature of the process and the type of raw material (particularly pigment) which is being used. This in turn affects their suitability for incineration. Some typical data are shown in Table 10.7.

There is a critical organic content required for incineration without the need for additional fuel, and Figure 10.5 shows the relationship between organic content and combustibility.

**Table 10.7**   *Paper and board sludge: production and composition.*
(Source: J.C. Roberts and P.W.W. Kirk, 'Paper Mill Sludge Production in the NW of England', *Environ. Technol. Lett.*, 1980, 1, 474–483).

| Dry sludge (kg/ton of production) | Solid content (%) | Fibre content (%) | Ash content (%) |
|---|---|---|---|
| 3–12 | 15–30 | 15–70 | 5–46 |

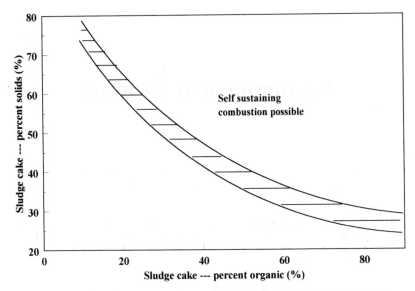

**Figure 10.5** *Relationship between organic content of sludges and their combustibility.* (Source: P.W.W. Kirk, 'Laboratory Evaluation and Postal Survey of Paper Mill Sludge in the NW of England', MSc Dissertation, University of Manchester, 1980, p. 28).

# Recommended Reading

'Wood Chemistry – Fundamentals and Applications', E. Sjostrom, Academic Press, San Diego, 1993, ISBN 0-12-647481-8.

'Handbook of Paper Science Volume 1 – The Raw Materials and Processing of Papermaking', ed. H.F. Rance, Elsevier, Amsterdam, 1980, ISBN 0-444-41778-8.

'Handbook of Paper Science Volume 2 – 'The Structure and Physical Properties of Paper', ed. H.F. Rance, Elsevier, Amsterdam, 1982, ISBN 0-444-41974-8.

'Chronology of the Origin and Progress of Paper and Paper Making', J. Munsell, Garland (USA), 1980, ISBN 0-8240-3878-9.

'Cell Wall Mechanics of Tracheids', R.E. Mark, Yale University Press, 1967.

'Pulp and Paper Chemistry and Technology' 3rd Edition, Volume 1, ed. J. Casey, J. Wiley & Sons, 1980, ISBN 0-471-03175-5.

'Lignin Biodegradation: Microbiology, Chemistry and Potential Applications', Volume 1, J. Kent Kirk, T. Higuchi and H.M. Chang; CRC Press, Boca Raton, USA, 1980, ISBN 0-8493-5459-5.

'Paper – An Engineered Stochastic Structure', M. Deng and C.T.J. Dodson, Tappi Press, Atlanta, USA, 1994, ISBN 0-8985-2283-8.

'Handbook of Physical and Mechanical Testing of Paper and Paperboard', R.E. Mark, Volume 1, 1983, Volume 2, 1984, Marcel Dekker, New York, ISBN 0-8247-7052-8 (Vol. 2) and 0-8247-1871-2 (Vol. 1).

'Subfracture Mechanical Properties', G.A. Baum, in 'Products of Paper Making'; Transactions 10th Fundamental Research Symposium, Volume 1, PIRA International, 1993, ISBN 1-85802-053-0, pp. 1–126.

'Paper Chemistry', ed. J.C. Roberts, Blackie, Glasgow, 1991, ISBN 0-216-92909-1.

'Cellulose Chemistry and its Applications', ed. T.P. Nevell and S.H. Zeronian, Ellis Horwood, Chichester, 1985, ISBN 0-85312-463-9.

'The Beating of Chemical Pulps – the Action and the Effects', D.H. Page in 'Transactions of the 9th Fundamental Research Symposium', ed. C.F. Baker and V. Punton, Volume 1, pp. 1–38, Mechanical Engineering Publications Ltd, London, 1989, ISBN 0-85298-706-4.

'Paper Chemistry – An Introduction', D. Eklund and T. Lindstrom, DT Paper Science Publications, Grankulla, Finland, 1991, ISBN 952-90-3606-X.

'Some Fundamental Chemical Aspects of Paper Forming', T. Lindstrom, in 'Transactions of the 9th Fundamental Research Symposium', ed. C.F. Baker and V. Punton, Volume 1, pp. 311–412, Mechanical Engineering Publications Ltd, London, 1989, ISBN 0-85298-706-4.

'The Coating Processes', Tappi Press, USA, 1993, ISBN 0-89852-266-8.

'Technology of Paper Recycling', ed. R.J. McKinney, Blackie, Glasgow, 1995, ISBN 0-7514-0017-3.

'Environmental Issues in the Pulp and Paper Industries – A Literature Review', N. Kirkpatrick, PIRA International, 1991, ISBN 0-90279-960-6.

'Handbook for Pulp and Paper Technologists', G.A. Smook, Angus Wilde Publications, Vancouver, 1992, ISBN 0-9694-6281-6.

'Principles of Wet End Chemistry' W.E. Scott, 1996. Tappi Press, Atlanta, USA, ISBN 0-89852-286-2.

# Subject Index

Abietic acid, 25, 126, 173
Absorption, 74
  of liquids by paper, 67–68
  of water by cellulose, 75–79, 80
*Acetobacter xylinum*, cellulose
  produced by, 21
Acidic pulp, *see* Sulfite pulp
Acidic pulping, 38–42
  and carbohydrate degradation,
    46–47, 49
  and solubilisation of organic
    extractives, 24–25
  sulfite, *see* Sulfite pulping
Adsorption, 74
  heat of, 75–76
  of polyelectrolytes, 100–108,
    113–117
Adsorption isotherm
  and retention mechanisms, 113,
    116
  for water, 73–74, 76–78
Air permeability, effects of recycling,
  155
Air-knife coating machine, 148–149
Algal celluloses, 21, 55
Alginates, as surface size, 146–147,
  148
Alkaline pulp, *see* Kraft pulp
Alkaline pulping, 42–44
  and carbohydrate degradation,
    45–46, 47
  and solubilisation of organic
    extractives, 24–25

Alkenyl succinyl anhydrides (ASAs),
  125, 128, 129, 130–131, 132,
  147–148
Alkyl ketene dimers (AKDs), 125,
  128–131, 147–148
Alum, 123
  use in internal sizing, 125,
    126–127, 128
  *see also* Aluminium sulfate
Alumina, 95
Aluminium sulfate
  as flocculant and retention aid,
    109–110, 113
  pH, 109, 125
  role in sizing, 127
  use with wet strength agents, 135
  *see also* Alum
Amylopectin, 145, 146, 147
Amylose, 145, 146, 147
Angiosperm, 12
  *see also* Hardwood
Anionicity, 100
Anisotropy of paper, 57
  and mechanical strength, 59–61
Arabinogalactan, 23
Arabino-(4-*O*-methylglucurono)-
  xylans, 23
Areal mass density distribution, *see*
  Mass density distribution

Bacteria
  celluloses from, 21, 55, 59